Praise for

THE BRAKE INDUSTRY HANDBOOK

"From OE program dynamics to aftermarket strategy to the technology shifts reshaping our industry, The Brake Industry Handbook covers the full landscape with the kind of insider detail you won't find anywhere else. With ever-changing dynamics of our industry I see a volume 2 in the future."

—**Frank Oliveto**, President,
Friction Materials Standards Institute (FMSI)

"Brian Hagman has spent decades at the heart of the brake industry and is widely regarded as one of its most trusted authorities. His ability to connect the technical detail with the commercial realities is rare. This handbook will become an indispensable reference for professionals across the global braking market."

—**Lloyd Ker**, CEO, Orthene Chemicals

"Everything I've learned about the brake industry over the last 15 years is in this book—and much more. Time is the only truly valuable thing in life, and Brian's masterpiece would have saved me an incredible amount of hours and energy had I read it back then. This will be the go-to resource for the next generation of professionals: if they make the right decision and read it carefully, they'll already be halfway to becoming best-in-class among the brake industry's most qualified experts."

—**José Luis Ciordia**, CEO, NUADI Group

"Brian does a great job breaking down the complexities of the braking industry. From the basics of how brakes operate, to formulations, to the commercial landscape, he provides a comprehensive yet accessible guide for anyone looking to understand this fascinating industry. A valuable resource for newcomers and veterans alike."

—**Greg Vyletel**, Ph.D., VP of Engineering,
PowerStop & Chair, SAE Brake Colloquium & Exhibition

THE BRAKE INDUSTRY HANDBOOK

THE BRAKE INDUSTRY HANDBOOK

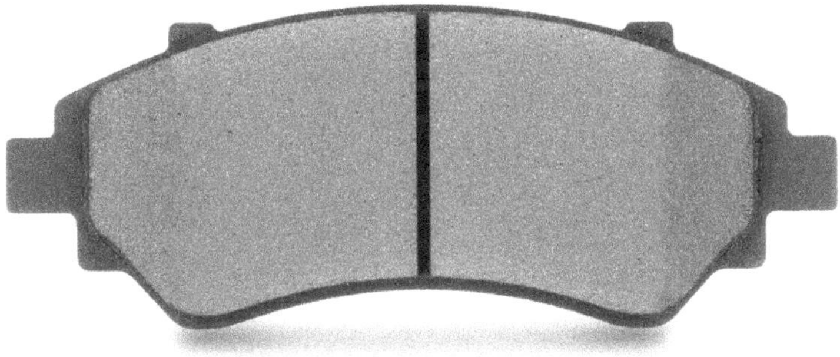

AN **INSIDER'S GUIDE** TO MARKETS, TECHNOLOGY, AND CAREERS

BRIAN **HAGMAN**

Founder of *The BRAKE Report* & Hagman Search

The Brake Industry Handbook: An Insider's Guide to Markets, Technology, and Careers

This book is provided for informational purposes only and reflects the author's opinions and industry experience. It does not constitute legal, financial, employment, or engineering advice. The author and publisher disclaim any liability for decisions made in reliance on its contents.

Industry data and market conditions described are current as of late 2025 and early 2026. Readers are encouraged to verify current information for decisions that depend on it.

Portions of this manuscript were developed with the assistance of AI-based research and drafting tools. See "A Note on Sources" for details.

For information about this title, contact the publisher:

Hagman Media
Elizabethtown, Kentucky, USA
www.hagmanmedia.com

ISBNs:
979-8-9951425-0-8 (hardcover)
979-8-99 51425-1-5 (softcover)
979-8-9951425-2-2 (eBook)

First Edition, 2026

Printed in the United States of America

Cover and Interior design: 1106 Design

To all those in the brake industry—past, present, and future. Your work is rarely seen, but it is felt every time a vehicle comes safely to a stop.

CONTENTS

FOREWORD

I've spent a career testing brakes. For more than five decades, through the eyes of Link Engineering Company, I have watched friction materials evolve, electronic systems transform vehicle dynamics, and safety standards push the entire industry toward capabilities that would have seemed impossible when I started. I've seen drum brakes give way to discs, mechanical systems yield to hydraulics and then to electronics, and simple anti-lock systems mature into sophisticated stability control. Through it all, two things have remained constant: there is never a dull moment in the Brake Industry and brakes don't get the attention they deserve.

That's not a complaint—it's simply the nature of the work. When brakes perform flawlessly, no one notices. A driver completes a panic stop and continues on their way, perhaps shaken but safe, rarely thinking about the engineering that made survival possible. The friction material that maintained its coefficient under extreme heat. The hydraulic system that distributed force precisely across four wheels. The ABS that pulsed dozens of times per second to prevent lockup. All of it invisible. All of it essential. Those of us in this industry take quiet pride in that invisibility—it means we're doing our jobs.

This invisibility extends to the industry itself. Automotive journalists write endlessly about horsepower and acceleration, styling and infotainment. The brake system—the one component that determines whether a vehicle can actually stop, and does everything the engine does in reverse—receives a passing mention at best. Business coverage follows the same pattern: endless analysis of electric vehicle manufacturers and battery suppliers, while the companies engineering the braking systems for those vehicles operate in relative obscurity. Even within the automotive supply chain, brakes remain underappreciated.

It was this reality that led Dr. Marv Weintraub of Ford Research Laboratories and me to establish the SAE Brake Colloquium nearly fifty years ago—a venue where engineers and others across the brake community could exchange ideas, share new developments, and accelerate the communication of knowledge that ultimately improves the safety of this critical product. That gathering has since inspired similar conferences worldwide, including EuroBrake in Europe and AsiaBrake in Asia, and the industry has benefited enormously from the interactions these events foster. Brian Hagman's BRAKE Report has complemented this activity by keeping the entire community current on product developments, industry events, and staying connected. This handbook is a natural guide and resource to not only navigate, but to chart our best path forward.

Brian has done something no one else has attempted: he's written a comprehensive guide to an industry that most people don't know exists. Not a technical manual, we have plenty of those. Not a market research report priced for corporate budgets. Instead, he's produced something genuinely useful: a handbook that explains how the Brake Industry actually works—the business dynamics, the career paths, the unwritten rules that take years to learn along with technical details to explain how brakes function.

I've known Brian through his work with *The BRAKE Report* and have encountered his recruiting work through Hagman Search. What strikes me about his approach is the dual perspective he brings. As a journalist, he's tracked the industry's evolution—the technology shifts, the regulatory pressures, the consolidation, the emergence of electric vehicles and autonomous driving. As a recruiter, he's learned what companies actually value in their people, how careers develop, and what separates those who thrive from those who struggle. That combination is rare, and it shows in these pages.

The technical chapters provide solid grounding in how brakes work and why they're engineered the way they are. The business chapters reveal dynamics that even industry veterans may not fully appreciate—how OE programs really operate, why aftermarket economics differ so dramatically, what drives the consolidation reshaping our competitive landscape. The career chapters offer practical guidance that I wish had existed when young engineers asked me for advice over the years. I found myself nodding along, recognizing patterns I'd observed but never seen articulated so clearly.

This book will serve multiple audiences. Engineers entering the Brake Industry will find the context they need to understand where their work fits in the larger picture. Business professionals will gain the technical foundation to engage credibly with engineering teams. Executives will find strategic perspective on an industry in transformation. And anyone considering a career in brakes, or trying to advance one, will discover insights that typically take years of experience to accumulate.

The Brake Industry stands at an inflection point. Electronic systems are redefining what brakes can do. Electric vehicles are changing how energy is managed. Regulations on particulate emissions are forcing complete material reformulation. Autonomous driving will eventually demand braking systems of unprecedented reliability and redundancy. The companies and professionals who navigate this transition successfully will rise to the top and will shape transportation safety for generations to come.

Understanding this industry has never been more critical, or more challenging. Brian Hagman has given us the guide and resource we will need to achieve this task going forward.

Roy H. Link
Chairman, Link Engineering Company

INTRODUCTION
WHY THIS BOOK EXISTS

No single resource explains how the brake industry actually works—the business dynamics, the career paths, the unwritten rules, the insider knowledge that separates those who thrive from those who struggle. Engineers can find technical papers. Executives can access market research reports costing thousands of dollars. But the practical wisdom that usually takes years of industry experience to accumulate? That hasn't existed in one place. This book is my attempt to fill that gap.

I've spent years embedded in the brake industry—not as an engineer or executive, but as something perhaps more unusual: a recruiter who also publishes the industry's leading trade publication. Through Hagman Search, I've reviewed thousands of résumés from brake industry professionals at every level, from fresh engineering graduates to C-suite executives. I've sat across the table from hiring managers at companies large and small, learning what they truly value—and what they say they value but don't. I've helped people navigate career transitions, negotiate compensation, survive acquisitions, and find their footing in an industry that can seem impenetrable to outsiders.

Through *The BRAKE Report*, I've tracked the industry's evolution since 2012—the technology shifts, the regulatory pressures, the consolidation, the rise of electric vehicles and autonomous driving. I've interviewed engineers, executives, and entrepreneurs. I've walked factory floors on three continents and attended more brake conferences than I can count. This dual role has given me a perspective that few others possess—seeing the industry through both its public stories and its confidential leadership decisions.

WHO THIS BOOK IS FOR

I wrote this book for several audiences, and I'd encourage you to use it in the way that serves you best:

If you're new to the brake industry—joining a brake company, working with brake suppliers, or evaluating the sector for investment—start at the beginning and read sequentially. Parts I and II build the foundation you'll need before the later chapters will pay off.

If you're an engineer or technical professional already working in braking, Parts I, II, and IV will deepen your understanding of the broader context around your work. The business chapters explain why decisions get made the way they do. The career chapters in Part V offer guidance you won't find in any employee handbook.

If you're a business professional—in sales, marketing, purchasing, or management—the technical chapters in Parts I and II provide the foundation you need to speak credibly with engineers and understand product conversations. Part II covers how OE and aftermarket business really work, and Part VI provides deeper dives into aftermarket strategy and electronic brake systems.

If you're considering a career in the brake industry, start with the career chapters in Part V, then circle back to the technical and business content. Understanding what roles exist, what companies look for, and how careers develop will help you make informed decisions about where to focus your energy.

If you're a student or academic, this book offers a practical complement to theoretical education. The brake industry is a remarkable case study in materials science, manufacturing, global supply chains, and regulatory response—all in a sector that directly impacts public safety.

If you're an investor, consultant, or journalist trying to understand the brake industry quickly, the market chapters and company directory provide orientation, while the deeper chapters offer the nuance that separates superficial analysis from genuine insight.

The appendices at the end provide a glossary of industry terms, a directory of major companies, and a list of key industry resources and associations—useful for any reader as a quick reference.

HOW THIS BOOK IS ORGANIZED

This handbook is structured in six parts, each building on the last:

Part I: Foundations (Chapters 1–5) lays the groundwork—the history of braking innovation, how brake systems actually work, and the market and competitive landscape including how geography shapes the industry across different regions.

Part II: How the Industry Works (Chapters 6–12) takes you inside the business—how OE programs are won and managed, the quality ecosystem that governs every supplier, friction material science, how the aftermarket operates, the supply chain structure that connects it all, manufacturing operations, and the testing and validation processes that ensure safety.

Part III: Application Segments (Chapters 13–14) covers specialized markets—motorsports as both a proving ground and brand-building platform, and the distinct world of commercial vehicle braking with its different physics, different suppliers, and different career paths.

Part IV: Forces Shaping the Future (Chapters 15–17) examines what's changing—the regulatory pressures reshaping the industry, the impact of electric vehicles on braking, and the technology frontier from brake-by-wire to autonomous vehicle integration.

Part V: Your Place in the Industry (Chapters 18–27) is the career guide—types of brake companies, functional work streams, geographic hubs, company evaluation frameworks, interview preparation, compensation negotiation, onboarding strategies, and navigating the hidden job market.

Part VI: Industry Deep Dives (Chapters 28–29) provides comprehensive treatment of aftermarket competitive strategy and electronic brake system technology for readers who want full depth on these critical topics.

A NOTE ON SOURCES

This handbook reflects years of direct industry experience through *The BRAKE Report* and Hagman Search, supplemented by publicly available information. The material draws on: conversations with hundreds of brake industry professionals conducted through executive search engagements and editorial interviews; ongoing coverage through *The BRAKE Report* since 2012, including factory visits, conference reporting, and executive conversations across North America, Europe, and Asia; annual attendance at the SAE Brake Colloquium and its proceedings and technical papers; regulatory documents from NHTSA, the European Commission, and their international counterparts; SEC filings and public financial disclosures from publicly traded brake suppliers; company press releases, annual reports, and investor communications; and trade media coverage across the automotive supplier industry. AI-based research and drafting tools assisted with structure, editing, and verification throughout the writing process.

A few important notes about the numbers. Market-size estimates vary meaningfully depending on whether analysts measure by revenue, units, or program count, and on where they draw the boundary between "brake system" and adjacent chassis content. Where I cite a range (for example, "$25–55 billion annually" for the global brake system market), the range reflects these definitional differences rather than uncertainty. Growth rates, market-share figures, and margin ranges are directional industry estimates drawn from multiple sources; in a rapidly consolidating industry with shifting regional dynamics, these will shift. Where I cite specific figures, they represent my best read of the available data as of late 2025 and early 2026.

Some of the most valuable material in this handbook—the unwritten rules, the career dynamics, the cultural differences between regional supply chains, the way hiring decisions actually get made—comes from my direct experience and from industry professionals who shared their knowledge in confidence. Where I've drawn on those conversations, I've preserved the insight while protecting sources. That's a deliberate choice consistent with how this industry operates: much of what matters travels through trusted relationships, not published reports.

I welcome feedback. The brake industry continues to evolve, and future editions of this handbook will evolve with it.

PART I
FOUNDATIONS

1.
THE RACE THAT CHANGED EVERYTHING

If you want to understand why the brake industry matters, start with a single race: the 1953 24 Hours of Le Mans.

Jaguar showed up with something nobody had seen before on a production-based race car: disc brakes. Their competitors were still using drum brakes—the same basic technology that had been around for decades. The Jaguar C-Types weren't the fastest cars on the straight, but they could brake later, harder, and more consistently than anything else on the track.

They dominated. First and second place, with the winning car setting a distance record that stood for years.

The motorsport world took notice. Within a few years, disc brakes became standard in racing. Within a decade, they started appearing on production vehicles. Today, disc brakes are so ubiquitous that most drivers don't realize there was ever another option.

This pattern—innovation born from competition, refined through engineering, and eventually made available for everyday drivers—has repeated throughout the history of braking. It's why I often tell people that the brake industry is really a story about the relentless pursuit of "better."

FROM WOODEN BLOCKS TO ELECTRONIC PRECISION

The earliest brakes were simple: a wooden block pressed against a wheel. If you've ever seen a horse-drawn carriage, you've seen the concept. It worked fine when top speeds were measured in single digits.

As vehicles got faster, those crude methods became dangerous. In the 1890s, a British engineer named Herbert Frood recognized the problem and began developing purpose-built friction materials from his workshop in Chapel-en-le-Frith, England—work that led to the founding of Ferodo and effectively launched the modern friction material industry. His original shop still stands on what is now the Tenneco campus, a quiet reminder that this global industry traces back to one man's workshop in the English countryside. Mechanical systems with levers and cables gave way to hydraulic brakes in the 1920s—a genuine revolution that allowed force to be multiplied and transmitted efficiently. Drum brakes became the standard, enclosing the braking mechanism and providing decent stopping power for the era.

Then came the disc brake breakthrough of the 1950s, followed by anti-lock braking systems (ABS) in the 1970s. ABS was initially met with skepticism—I've talked to engineers who remember the debates about whether "pumping" brakes electronically was really better than letting skilled drivers do it themselves. The data settled that argument definitively: ABS saves lives.

Today, we're in the era of electronic integration. Brake-by-wire systems, electronic stability control, automatic emergency braking—these technologies would have seemed like science fiction to the engineers who first bolted disc brakes onto that Jaguar. And yet, the fundamental challenge remains the same: convert kinetic energy into heat, quickly and reliably, without killing anyone.

Simple to state. Extraordinarily difficult to execute.

KEY TAKEAWAYS

- Jaguar's disc brakes at the 1953 Le Mans marked a turning point—proving that better braking technology creates competitive advantage.
- Disc brakes succeeded because they offered better heat dissipation and more consistent performance than drums, not just more stopping power.
- This race illustrates a pattern that continues today: brake innovation often emerges from motorsports before reaching production vehicles.
- The brake industry may lack glamour, but it has consistently delivered critical safety advances that save lives daily.

2.
HOW BRAKES ACTUALLY WORK

When people outside our industry ask me what makes braking interesting, I often start with this:

"Do you know what happens when you press your brake pedal?"

The typical answer: "The brakes slow down the wheels."

That's true, but it misses the fascinating physics. When you brake, you're converting kinetic energy—the energy of your moving vehicle—into thermal energy. Heat. A lot of heat.

A single hard stop from highway speed can push your brake rotors past 500°F. In racing applications, brake temperatures regularly exceed 1,000°F. That's hot enough to glow visibly. If you've ever watched motorsports at night, you've probably seen the eerie orange glow of overheated brakes.

This is why brake engineering is fundamentally materials science. Every component must perform consistently across an enormous temperature range, from freezing winter mornings to the heat generated by aggressive driving. Get the material formulation wrong, and brakes either wear too quickly, make excessive noise, don't stop effectively, or fail catastrophically.

THE ANATOMY OF A STOP

Your foot pushes the brake pedal, which activates the master cylinder—the heart of the hydraulic system. The master cylinder converts your mechanical push into hydraulic pressure, multiplying your force significantly. That pressure travels through brake lines and hoses filled with brake fluid, a specialized liquid engineered to remain stable at extreme temperatures and resist compression.

In a disc brake system, this hydraulic pressure reaches the calipers—the clamping mechanisms that straddle the rotors. Inside each caliper are pistons that push the brake pads against the spinning rotor (also called a disc). The friction between pad and rotor generates the heat that converts your kinetic energy into thermal energy, slowing your vehicle.

Drum brakes work on the same principle but with different geometry: brake shoes press outward against the inner surface of a spinning drum. They're still found on some vehicles, typically on rear wheels where braking demands are lower, but disc brakes dominate the modern market—accounting for the large majority of OEM installations globally.

Every single component in this chain must work flawlessly, every time, under conditions ranging from gentle parking lot stops to panic braking on a rain-slicked highway. When I talk to brake engineers, this is what keeps them focused: the knowledge that their work will be tested in life-or-death scenarios they can't fully predict.

ROTOR ENGINEERING

The brake rotor seems simple—a metal disc that spins with the wheel. But rotor engineering involves sophisticated trade-offs between weight, thermal capacity, cooling efficiency, and cost.

Solid vs. Vented Rotors

Solid rotors are exactly what they sound like—a single solid disc of cast iron. They're lighter, cheaper to manufacture, and perfectly adequate for

FIGURE 1: Disc Brake Assembly

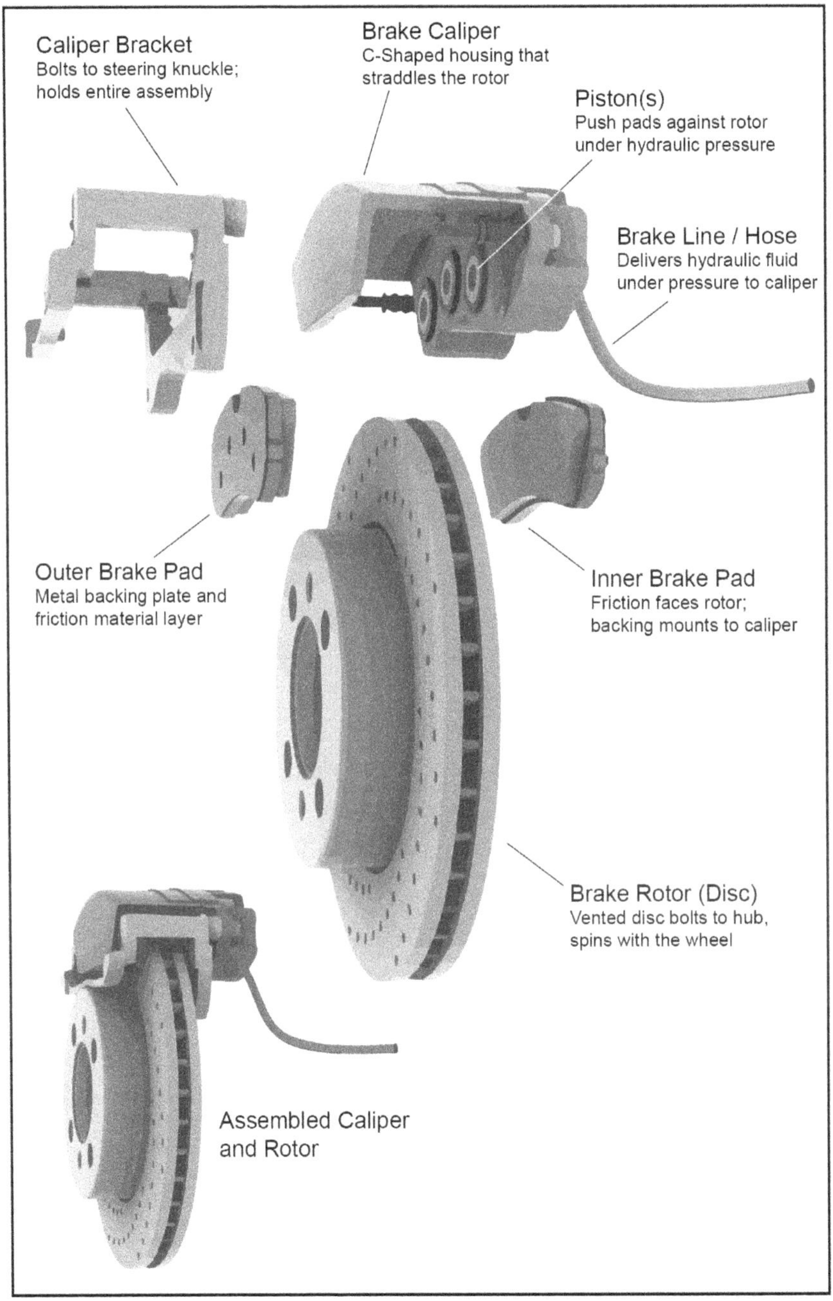

low-demand applications like rear brakes on lighter vehicles. You'll find them where braking loads are modest and heat buildup isn't a primary concern.

Vented rotors feature internal cooling vanes between two friction surfaces. These vanes act as a centrifugal pump—as the rotor spins, air is drawn in through the center and expelled at the outer edge, dramatically improving heat dissipation. A vented rotor can dissipate heat substantially more effectively than a solid rotor of the same diameter.

The vane design itself varies significantly. Straight radial vanes are simplest and cheapest. Curved or "pillar" vanes improve airflow but add manufacturing complexity. Some high-performance rotors use directional vane patterns optimized for rotation direction, requiring left-hand and right-hand specific parts.

Two-Piece Rotor Designs

Premium and performance applications often use two-piece rotors that separate the friction ring (the part the pads contact) from the hat (the center section that bolts to the hub). This design offers several advantages:

The friction ring can be made from optimized iron while the hat uses lighter aluminum, reducing unsprung weight. The floating connection between ring and hat allows differential thermal expansion—the friction ring can expand as it heats without inducing thickness variation in the mounting surface. The design theoretically allows ring-only replacement, though in practice most owners replace the full assembly—replacement rings alone are not widely available as service parts. Some aluminum hats use riveted connections (as on certain BMW models) while others are bolted (such as on the Charger Hellcat).

The trade-off is cost and complexity. Two-piece rotors can cost five to ten times more than one-piece equivalents, limiting their use to performance vehicles and motorsports.

Most brake rotors are gray cast iron, chosen for its exceptional balance of thermal conductivity, damping, and machinability at acceptable cost. The specific iron composition—particularly carbon content and graphite flake structure—significantly affects rotor performance. For the full technical breakdown of rotor metallurgy, including carbon content ranges, graphite morphology, and high-carbon rotor considerations, see Appendix E: Technical Deep Dives.

Thermal Considerations

A rotor must absorb enormous amounts of heat without failing. A single emergency stop from highway speed can raise rotor temperature by 500°F or more. Repeated hard stops—descending a mountain grade, for example—can push temperatures past 1,000°F, with fade testing sequences reaching 1,500°F (over 800°C).

At these temperatures, several failure modes threaten:

Thermal cracking is typically the result of thermal mechanical fatigue—repeated thermal cycling under load that progressively weakens the material. When a rotor is mated with an incompatible friction material, thermal elastic instabilities (commonly called hot spotting) can develop, creating very high localized temperatures that accelerate the cracking process. It is a complex phenomenon. These cracks typically radiate from the hub or run circumferentially around the friction surface. Minor heat checking (small surface cracks) is normal and doesn't indicate failure. Deep cracks that compromise structural integrity require replacement.

Hot spots develop when localized areas of the rotor become harder than surrounding material due to thermal transformation. These hard spots wear differently, creating thickness variations that cause pedal pulsation—a condition widely but incorrectly attributed to "warped rotors." Rotors do not actually warp. Even the slightest thickness variation will produce perceptible pulsation, and this variation—not warping—is what drivers experience.

Thermal judder results from thickness variation regardless of cause. Variation as small as 0.0005 inches (half a thousandth) can produce perceptible pulsation. This is why rotor runout and thickness variation specifications are so tight, and why proper torquing of wheel fasteners matters—uneven clamping force can create the very distortion it's meant to prevent.

Minimum Thickness and Why It Matters

Every rotor has a minimum thickness specification cast or stamped into it. This isn't arbitrary—it represents the point at which the rotor can no longer safely absorb and dissipate heat without risk of failure.

As rotors wear, they lose thermal mass. A rotor worn to minimum thickness may still stop the vehicle under normal conditions but lacks the heat capacity to handle repeated hard stops or sustained braking. Worse, a thin rotor flexes more under clamping pressure, potentially allowing the caliper pistons to extend beyond their design limits and increasing the risk of fluid overheating.

The minimum thickness also accounts for structural integrity—a rotor worn below minimum may crack or even separate under stress. There is also a critical tolerance stack-up to consider: at minimum rotor thickness with fully worn pads, the pad backing plate must still be retained by the caliper anchor bracket to maintain hydraulic pressure. If this stack-up is not maintained, the pistons can overextend and the pedal goes to the floor.

CALIPER DESIGN

The caliper's job is conceptually simple: squeeze the pads against the rotor with sufficient force to stop the vehicle. The engineering behind that squeezing, however, involves significant complexity.

Floating vs. Fixed Calipers

Floating (or sliding) calipers use a piston on only one side of the rotor. When hydraulic pressure is applied, the piston pushes the inboard pad against the rotor. The reaction force then pulls the caliper body (which "floats" on sliding pins) toward the rotor, pressing the outboard pad against the opposite face. The caliper effectively straddles the rotor and self-centers.

Fixed calipers have pistons on both sides of the rotor, pushing inboard and outboard pads simultaneously. The caliper itself is rigidly mounted and doesn't move.

Floating calipers dominate the market because they're simpler, lighter, and cheaper. They accommodate rotor thickness variation more easily and require less precise mounting. Their disadvantage is the sliding mechanism itself—the pins can corrode, seize, or bind, causing uneven pad wear and reduced braking effectiveness. Regular lubrication of slide pins is important maintenance that's often neglected.

Fixed calipers offer superior stiffness and more consistent clamping force, which translates to better pedal feel and more consistent performance under hard use. With pistons on both sides, they can achieve the same clamping force with smaller individual pistons, or greater total force with the same packaging. The Brembo calipers visible through the wheels of sports cars are typically fixed multi-piston designs.

The trade-offs are cost and packaging complexity. Most light vehicle fixed calipers are aluminum, which makes them lighter than their iron floating counterparts—but they are generally more expensive, require different wheel packaging space, and demand more precise manufacturing. They're common on performance and premium vehicles and increasingly found on mainstream platforms as automakers pursue better pedal feel and braking consistency.

Piston Configurations

Single-piston floating calipers handle the majority of passenger vehicles. The simplicity is hard to beat for cost-sensitive applications.

Multi-piston calipers (whether floating or fixed) distribute clamping force more evenly across the pad face. A four-piston caliper with two pistons per side can use smaller pistons than a two-piston design while achieving similar total force—and the distributed application reduces pad taper wear.

Six-piston and even eight-piston calipers appear on high-performance applications. The progression isn't just about raw clamping force—it's about how that force is distributed across increasingly large pads designed to handle extreme heat loads.

Opposed-piston fixed calipers (pistons directly opposing each other across the rotor) provide the most symmetrical loading. This design, pioneered by Brembo for motorsports and now common on performance vehicles, maximizes caliper stiffness while minimizing the flex that degrades pedal feel.

FIGURE 2: Floating vs. Fixed Caliper Design

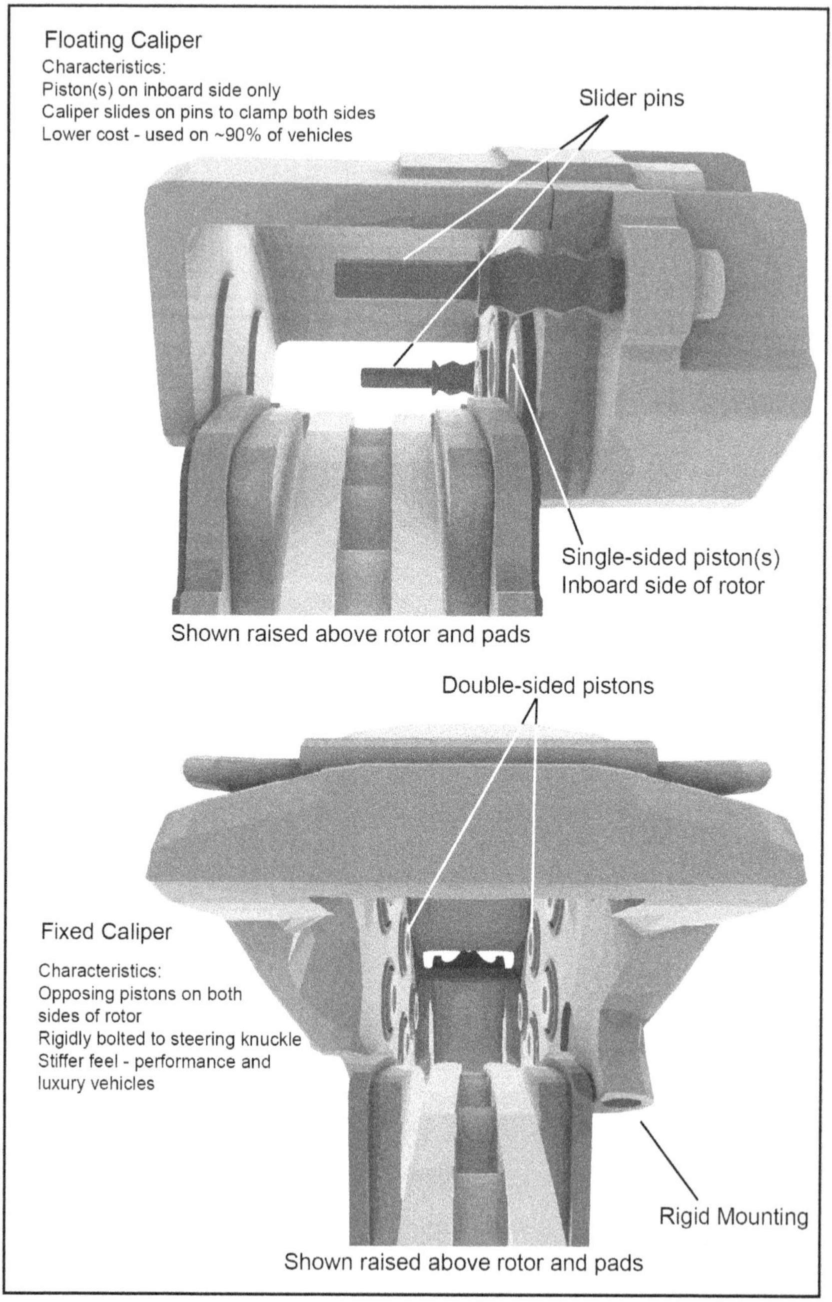

Caliper Materials

Iron calipers are common on trucks, SUVs, and heavy vehicles where strength and heat resistance matter more than weight. Cast iron is inexpensive and handles heat well, though it's heavy and prone to corrosion.

Aluminum calipers reduce unsprung weight significantly—important for ride quality, handling, and suspension performance. Aluminum's lower density (about one-third that of iron) means an aluminum caliper can weigh roughly half of an equivalent iron design. The trade-offs include thermal management and stiffness: aluminum expands more with heat, softens at lower temperatures, and can transfer more heat to the brake fluid. Aluminum is also not as stiff as nodular iron, so aluminum calipers require thicker cross-sections to achieve equivalent rigidity—which adds material and can offset some of the weight savings.

Modern aluminum calipers are typically made from high-strength alloys and designed with generous material in critical areas. Anodizing or other surface treatments improve corrosion resistance. Most passenger car calipers today are aluminum.

Caliper pistons are typically either steel or phenolic (plastic composite). Steel pistons are strong and conduct heat well; phenolic pistons insulate brake fluid from heat, reducing the risk of fluid boiling. Most modern calipers use phenolic pistons for their thermal advantages. For the detailed comparison, see Appendix E: Technical Deep Dives.

THE HYDRAULIC SYSTEM

The hydraulic system is the communication link between your foot and the wheel brakes. Understanding it explains why brake feel varies so much between vehicles.

Brake fluid is engineered to transmit force without compression across an enormous temperature range. The DOT classification system (DOT 3, 4, 5, and 5.1) grades fluids primarily by boiling point—critical because all glycol-based fluids absorb moisture over time, progressively lowering their boiling point. When brake fluid boils under hard use, the resulting vapor bubbles compress instead of transmitting force, causing sudden pedal loss. This is why brake fluid requires periodic replacement regardless of apparent condition. For the complete DOT specification breakdown and boiling point data, see Appendix E: Technical Deep Dives.

Master Cylinder Function

The master cylinder converts mechanical motion (your foot pushing the pedal) into hydraulic pressure. Modern vehicles use tandem master cylinders with two separate pistons creating independent pressure for front and rear circuits. If one circuit fails, the other continues to function—you'll have reduced braking ability but not total failure.

FIGURE 3: Hydraulic Brake System

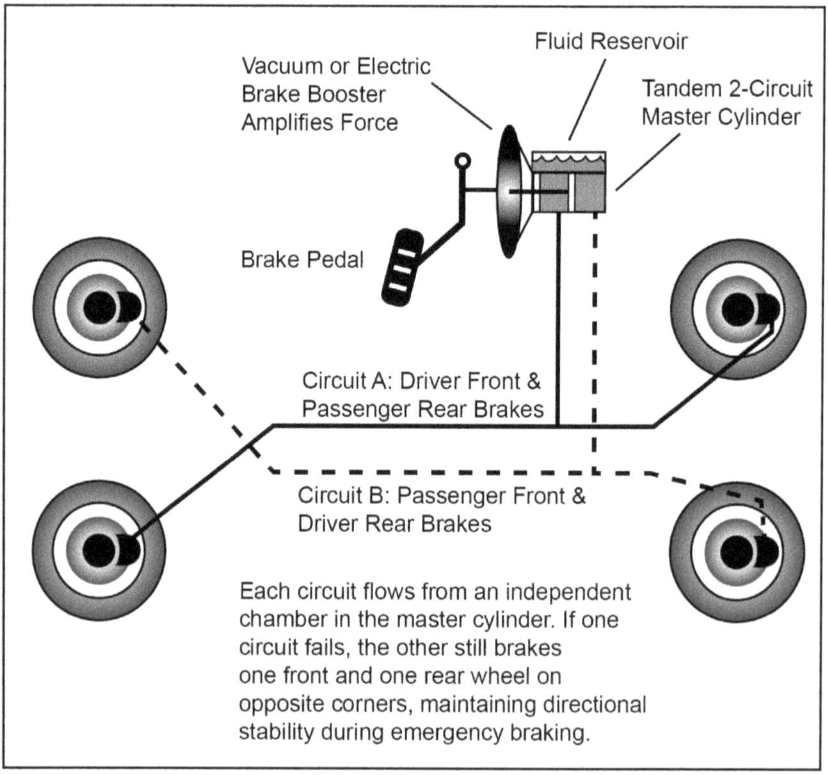

Master cylinder bore size matters. A larger bore moves more fluid volume per stroke but requires more pedal force to generate equivalent pressure. A smaller bore requires less force but moves less fluid—the pedal moves farther before brakes engage. Engineers balance these trade-offs to achieve desired pedal feel and travel.

Brake Bias and Proportioning

Under braking, weight transfers forward. The front brakes therefore do most of the work—typically 60–70% of total braking force even on modern vehicles. Electronic brake distribution (EBD) allows the rear brakes to contribute more effectively than fixed proportioning systems by adjusting rear pressure dynamically based on load and conditions, but the fundamental physics of weight transfer means the front always dominates. Brake systems are designed with this bias built in, using larger front rotors, larger calipers, or both.

But optimal bias changes with load. An empty pickup truck needs different front/rear balance than one carrying a heavy payload. Proportioning valves, load-sensing valves, or ABS systems adjust rear brake pressure based on conditions to prevent premature rear wheel lockup while maximizing total braking effectiveness.

THE PHYSICS OF STOPPING

Understanding the physics helps explain why brakes are sized the way they are—and why upgrading brakes isn't as simple as bolting on bigger rotors.

Brake torque—the force that slows the wheel—is a function of clamping force, coefficient of friction, and the effective radius where braking force acts. Larger rotors increase effective radius, which is why bigger brakes improve leverage rather than just adding thermal mass. For the full torque calculation and physics, see Appendix E: Technical Deep Dives.

Energy and Heat

Brakes convert kinetic energy to heat. The kinetic energy of a moving vehicle equals ½ × mass × velocity^2.

That squared velocity term explains why stopping from 60 mph doesn't require twice the braking of stopping from 30 mph—it requires four times the braking. Double your speed and you quadruple the energy that must be dissipated as heat.

A single hard stop from 60 mph in a typical passenger car converts roughly 500,000 joules (about 120,000 calories) to heat—enough to raise the temperature of the brake rotors by several hundred degrees. Repeated stops accumulate heat faster than it can dissipate, which is why sustained hard use (mountain descents, track driving) creates conditions far more demanding than normal driving.

Why Brakes Are Sized for Sustained Use

Brake systems aren't designed for single panic stops—they're designed for sustained braking scenarios like descending a long mountain grade. Brake sizing is a true optimization balancing vehicle mass (most brake systems are shared across several different vehicle models), wheel size constraints, desired ride quality and mass expectations, and the thermal management demands of the worst-case duty cycle. A loaded vehicle descending a 6% grade for several miles at 40 mph must continuously dissipate thousands of watts of heat.

This explains why simply "upgrading" to larger brakes doesn't always improve stopping distance. The stock brakes can likely lock the wheels in a single panic stop—there's no friction limitation. The upgrade improves thermal capacity and sustained performance, not single-stop capability.

Truly improving single-stop performance requires better tires. The brake system can always generate enough torque to overwhelm tire grip; it's the tires that determine maximum deceleration. This is why racing teams obsess over tire compounds and temperatures as much as brake systems.

KEY TAKEAWAYS

- Braking is fundamentally about converting kinetic energy to thermal energy—a single hard stop from highway speed can push rotor temperatures past 500°F.
- Disc brakes dominate the market (the large majority of OEM installations) because they dissipate heat more effectively and deliver more consistent performance than drum brakes.
- Rotor engineering involves trade-offs between weight, thermal capacity, and cost—vented rotors dissipate heat substantially more effectively than solid rotors.
- Floating calipers dominate due to lower cost and simpler packaging, while fixed multi-piston calipers offer superior stiffness and pedal feel for performance applications.
- Brake fluid is hygroscopic—it absorbs moisture over time, lowering its boiling point and potentially causing brake fade under hard use.

3.
THE MARKET LANDSCAPE

Let's talk numbers. The global automotive brake system market is substantial—valued at roughly $25–55 billion annually depending on how you define the scope (complete systems vs. components, passenger vehicles vs. all vehicles). Most analysts project steady growth in the 4–6% range annually through the next decade, driven by rising vehicle production, increasing safety requirements, and the complexity demands of electric and autonomous vehicles.

But those headline figures mask the real story, which is about market structure and where value flows.

OE VS. AFTERMARKET: TWO DIFFERENT BUSINESSES

The brake industry operates in two fundamentally different channels: original equipment (OE) and aftermarket.

Original Equipment involves supplying brake systems and components to vehicle manufacturers for installation on new vehicles. This is a high-volume, relationship-intensive business where contracts are won through engineering capability, quality consistency, and cost competitiveness.

OE represents the vast majority of brake system revenue globally—it's where the major players make their mark.

Aftermarket involves replacement parts sold after the vehicle leaves the factory. This market is driven by wear-and-tear (brake pads and rotors are consumables that require periodic replacement), repair needs, and performance upgrades. The aftermarket is more fragmented, with a broader range of players including regional specialists and value-oriented manufacturers.

The dynamics are different in each channel. OE suppliers need massive engineering resources, global footprints, and the ability to co-develop systems with automakers years before a vehicle launches. Aftermarket players need distribution networks, brand recognition among repair shops and consumers, and competitive pricing.

Some companies play in both channels—Brembo, Akebono, TMD Friction, and Frasle Mobility, for example, are strong in both OE and aftermarket. Others specialize. Understanding which channel a company focuses on tells you a lot about their business model and competitive positioning.

REGIONAL DYNAMICS

Geography matters enormously in the brake industry. Asia-Pacific—led by China, Japan, and South Korea—is the largest regional market for brake systems, reflecting the region's dominance in vehicle manufacturing. The sheer scale of Chinese vehicle production has made it the center of gravity for the industry.

Europe remains critical for premium positioning and technology leadership. Companies like Brembo, AUMOVIO (formerly Continental's automotive division), and other major German suppliers have shaped industry standards and continue to drive innovation, particularly in high-performance and electric vehicle applications. Euro 7 emissions regulations are also forcing technology advancement in ways that will ripple globally.

North America represents mature but substantial demand. The U.S. market alone is worth roughly $9 billion annually, with particular strength in light trucks and SUVs that require robust braking systems. The aftermarket is especially significant here, supported by the large installed base of vehicles and a strong DIY culture.

MARKET SEGMENTATION

Beyond OE/aftermarket and geography, the market segments by vehicle type and component:

By vehicle: Passenger cars dominate revenue—representing the large majority of the market—but commercial vehicles, heavy trucks, and buses represent distinct segments with different requirements and different suppliers. The heavy-duty segment alone—covering trucks, buses, and trailers—is a substantial market with specialized players like Knorr-Bremse, WABCO (now part of ZF), Bendix, and Cummins Drivetrain and Braking Systems (formerly Meritor).

By component: Brake pads and friction materials represent the largest component category due to their replacement cycles—global production exceeds 4 billion units annually. (The brake block, or lined shoe, also remains dominant in the commercial vehicle segment.) But the fastest growth is in electronic components: control units, sensors, and actuators are climbing at double-digit rates annually as value migrates toward software-centric systems.

KEY TAKEAWAYS

- The global brake system market is valued at $25–55 billion annually, with steady 4–6% growth projected through the next decade.
- OE (Original Equipment) represents the vast majority of brake system revenue; aftermarket is more fragmented with different competitive dynamics.

- Asia-Pacific is the largest regional market for brake systems, reflecting its dominance in vehicle manufacturing.
- The fastest growth is in electronic components—control units, sensors, and actuators are climbing at double-digit rates annually as value migrates toward software-centric systems.

4.
THE COMPETITIVE LANDSCAPE

The brake industry features a mix of global giants and regional specialists. At the top, a handful of players command significant market share; the top five suppliers are typically estimated to account for roughly half of global OEM partnerships, though exact figures vary depending on whether you measure by revenue, units, or program count. But beneath that, hundreds of companies serve various niches, regions, and applications.

Let me walk you through the major players:

THE GLOBAL LEADERS

Brembo (Italy) is perhaps the most recognized name in braking, synonymous with high-performance applications. Founded in 1961, Brembo supplies braking systems to premium automakers, motorsports teams, and the performance aftermarket. Their red calipers are visible on everything from Ferraris to police pursuit vehicles. Brembo has invested heavily in carbon-ceramic technology and recently launched Sensify, a software-driven braking system that modulates each wheel independently. They've also been expanding their position in electric vehicles, where their expertise in managing high heat loads from regenerative braking integration is valuable.

Robert Bosch (Germany) brings massive scale and integration capability. As the world's largest automotive supplier, Bosch's braking division benefits from synergies with their broader electronics, sensors, and software businesses. Their iBooster and now eBooster 2.0 systems integrate braking with regenerative energy recovery for EVs. Bosch's strength is in complete systems integration—they can deliver braking as part of a comprehensive chassis or ADAS solution.

AUMOVIO (Germany), formerly Continental's automotive group sector, became an independent publicly traded company in September 2025. AUMOVIO reports annual sales of roughly €19.4 billion and approximately 92,000 employees, making it one of the largest automotive technology companies globally. The company's brake portfolio includes electronic braking systems, brake-by-wire technology, and ADAS integration—they produced over 1 million electronic brake systems in India alone in 2024.

ZF Friedrichshafen (Germany) expanded dramatically in braking through its 2020 acquisition of WABCO, the commercial vehicle braking specialist. ZF now spans passenger vehicles, commercial trucks, and off-highway applications. Their strength is in integrating braking with broader chassis systems—steering, suspension, and driveline—to create coordinated vehicle control.

Akebono (Japan) has built its reputation on quiet, low-dust ceramic brake pads and a strong presence in North American and Asian OE markets. They've been leaders in developing copper-free formulations to meet environmental regulations. Akebono supplies both OE and aftermarket channels and has maintained strong relationships with Japanese automakers while expanding globally.

Aisin (Japan) operates within the Toyota ecosystem but serves other automakers as well. Their braking business benefits from Toyota's massive production volumes and focus on reliability. Aisin has been investing in brake-by-wire and integrated chassis systems as vehicles become more electrified.

ITT Motion Technologies (Italy, parent: ITT Inc.) is a leading European OE brake pad supplier with deep relationships among premium automakers, particularly German luxury brands. Headquartered in Barge, Italy, ITT Friction Technologies operates manufacturing sites across Italy, the Czech Republic, Mexico, and the U.S., and recently invested over €50 million to expand its Termoli, Italy facility for high-performance brake pad production. As part of NYSE-listed ITT Inc., the Motion Technologies segment also includes shock absorber and damping businesses (KONI), giving ITT broader chassis exposure than pure-play friction specialists.

COMMERCIAL VEHICLE SPECIALISTS

The commercial vehicle segment has its own cast of characters, with different technology requirements (air brakes vs. hydraulic brakes) and different customers.

Knorr-Bremse (Germany) is the global leader in commercial vehicle and rail braking systems. They dominate the heavy truck and trailer segment, with particular strength in Europe and growing presence in Asia. Knorr-Bremse has been expanding into driver assistance systems and electronic controls, positioning braking as part of broader commercial vehicle safety.

Bendix Commercial Vehicle Systems (USA) is the North American leader in commercial vehicle braking, with a long corporate heritage and now owned by Knorr-Bremse. They've recently streamlined their portfolio to focus on core braking competencies, launching new products like the ADB22X-LT air disc brake optimized for lighter-duty commercial applications.

Cummins Drivetrain and Braking Systems (USA), formerly Meritor, remains a major force in commercial vehicle braking with a significant global presence in both foundation brake products and aftermarket support. Meritor was acquired by Cummins Inc. in 2022 and became a business unit within Cummins' Components segment, rebranded as Cummins Drivetrain and Braking Systems in 2025. The Meritor name is retained as the aftermarket brand for the group's brake and drivetrain components.

Haldex (Sweden), acquired by SAF-Holland in 2022, focuses on brake adjusters, air suspension, and other commercial vehicle systems. They've carved out strong positions in specific component categories where they can achieve technology leadership.

REGIONAL AND EMERGING PLAYERS

Beyond the global giants, numerous regional players serve local markets or specific niches:

Chinese manufacturers have grown dramatically, moving from serving domestic automakers to competing credibly for global OE programs. The number of Chinese brake suppliers has grown significantly over the past two decades, and many have expanded beyond China's borders. Chapter 5 explores the specific dynamics—compressed development timelines, EV technology leadership, and the key players driving this shift.

Indian manufacturers—including Brakes India, Sundaram Brake Linings, Rane Brake Linings, ASK Automotive, and Masu Brakes—are becoming significant forces in global supply chains as OEMs seek diversification. Chapter 5 covers India's growth trajectory and the dynamics shaping these companies' expansion.

Latin American manufacturers led by Frasle Mobility (Brazil)—the region's largest friction producer, with operations now spanning the Americas, Europe, and Asia after a string of acquisitions including Federal-Mogul's Sorocaba plant, Fremax, Juratek, and most recently the Mexican aftermarket group Dacomsa—have moved well beyond their regional origins.

Aftermarket specialists like MAT Holdings (USA), PowerStop (USA), Dynamic Friction Company (USA), EBC Brakes (UK), Wilwood Engineering (USA), and numerous others focus specifically on replacement parts and performance upgrades rather than OE supply.

The competitive dynamics are shifting. Electrification and electronic integration are raising the bar for engineering capability. Companies that can't develop software-centric braking solutions risk being commoditized or acquired. At the same time, the transition to EVs opens opportunities for new entrants who aren't burdened by legacy hydraulic system investments.

KEY TAKEAWAYS

- The top five brake suppliers are typically estimated to account for roughly half of global OEM partnerships, but hundreds of companies serve various niches below them.
- Brembo dominates high-performance; Bosch and AUMOVIO lead in electronics; Knorr-Bremse and ZF lead commercial vehicles.
- Regional players—particularly in China and India—are increasingly competitive and moving beyond their home markets. Chapter 5 examines how geography shapes these competitive dynamics.
- The competitive landscape is consolidating—scale matters more as electronic content increases and R&D costs rise.

5.
THE GLOBAL BRAKE INDUSTRY: HOW GEOGRAPHY SHAPES EVERYTHING

Chapter 3 introduced the regional market sizes, and Chapter 4 covered the major players. But knowing that Asia-Pacific is the largest regional market doesn't tell you how doing business in Shanghai differs from doing business in Stuttgart or Detroit. The rules change depending on where you are—and anyone working at a company with global operations (which is most brake suppliers of any scale) needs to understand those differences.

This chapter won't try to be a comprehensive country-by-country guide. Instead, it covers the dynamics that most directly affect how people in the brake industry actually work.

CHINA: THE CENTER OF GRAVITY

China produces roughly 30 million vehicles per year—more than any other country by a wide margin. That sheer volume has made it the most important market for every global brake supplier, and the most competitive.

The landscape looks different here than in North America or Europe. Local Chinese brake companies—firms like Bethel Automotive Safety

Systems, Zhejiang Asia-Pacific Mechanical & Electronic (APG), and Wanxiang Qianchao—have grown rapidly and now compete credibly for OE programs that would have gone exclusively to multinationals a decade ago. They've invested heavily in engineering capability, particularly in electronic brake systems, and they benefit from cost structures and speed that Western suppliers struggle to match.

Relationships with Chinese automakers operate on a compressed timeline. Where a European OEM might run a three-year development cycle, Chinese OEMs—particularly the newer EV-focused brands like BYD, NIO, and Li Auto—often work on 18-to-24-month cycles. Suppliers who can't match that pace lose opportunities regardless of their technical capability.

The EV transition is more advanced in China than anywhere else, which has direct implications for brake suppliers. China leads global adoption of electro-hydraulic boost systems and brake-by-wire technology. Companies that develop these capabilities for the Chinese market often find themselves ahead when the same technologies roll out in Europe and North America.

For careers, China-facing roles are increasingly important at every major supplier. Mandarin language skills aren't required for most positions (English is the working language in most multinational brake companies' China operations), but cultural fluency—understanding how decisions get made, how relationships are built, and how negotiations work differently—is genuinely valuable.

EUROPE: WHERE REGULATION DRIVES INNOVATION

Europe's importance to the brake industry is disproportionate to its vehicle volume. It's where many of the technologies that eventually reach global markets are developed first, largely because European regulations push the envelope.

Euro 7 brake particle emissions requirements are the clearest example. Europe is the first major market to regulate non-exhaust particulate emissions from braking, and the standards taking effect are forcing material science innovation that will eventually reshape formulations worldwide. Suppliers who solve the Euro 7 challenge first gain a durable competitive advantage.

ECE R90 governs aftermarket brake components in Europe and creates a fundamentally different market structure than North America. Every aftermarket brake pad and rotor sold in Europe must be homologated—tested and certified to meet specific performance standards for each vehicle application. This creates a significant barrier to entry that keeps low-cost, unbranded products out of the European aftermarket in ways that simply don't apply in the U.S. It also means aftermarket margins in Europe tend to be healthier, but the cost of maintaining ECE R90 certifications across thousands of part numbers is substantial.

The German automakers—Volkswagen Group, BMW, Mercedes-Benz, and their associated brands—remain the most technically demanding brake customers in the world. Their specifications go well beyond regulatory minimums, particularly around NVH performance, and their engineering validation processes are exhaustive. Suppliers who can satisfy BMW's brake noise requirements can usually satisfy anyone's.

European brake engineering talent tends to be deeply specialized. Engineers at companies like AUMOVIO, Bosch, ZF, TMD Friction, and ITT Motion Technologies often spend entire careers in braking, developing expertise that's difficult to replicate. The engineering culture is more formal than in the U.S.—technical documentation standards are rigorous, and career progression tends to follow structured paths.

NORTH AMERICA: MATURE MARKET, EVOLVING DYNAMICS

North America is the market most readers of this book will know best, but a few dynamics are worth calling out explicitly because they shape career opportunities and competitive dynamics.

The light truck and SUV mix matters enormously for brake suppliers. North American consumers have shifted decisively toward larger vehicles, which require more robust braking systems—larger rotors, higher-capacity calipers, and friction materials tuned for heavier applications. This has been a tailwind for suppliers with strong light truck portfolios and a challenge for those optimized primarily for passenger cars.

The North American aftermarket is the most open in the world. Unlike Europe's ECE R90 regime, the U.S. has minimal regulatory barriers for aftermarket brake components. FMVSS 135 governs OE brake systems, but aftermarket replacements face no comparable homologation requirement. This openness enables a thriving independent aftermarket with hundreds of brands, but it also means quality variation is wider than in Europe. Understanding quality tiers (see Chapter 9) matters here more than anywhere.

Detroit's Big Three—General Motors, Ford, and Stellantis—remain critical customers, but their sourcing approach has evolved. They're increasingly willing to source from non-traditional suppliers, including Chinese and Indian companies, for certain components. At the same time, their engineering teams have become more technically capable at evaluating brake performance internally, making the customer relationship more technically rigorous than it was a decade ago.

Mexico has become a major brake manufacturing hub. Multiple global suppliers operate significant production facilities there, serving both the Mexican vehicle assembly industry and the U.S. aftermarket. For engineers and operations professionals, willingness to support Mexican manufacturing operations—whether through relocation or frequent travel—has become a common requirement in many roles.

INDIA: THE GROWTH STORY

India produces roughly 5 million vehicles annually and is projected to grow significantly. The market is dominated by price-sensitive segments—smaller vehicles with correspondingly smaller brake systems—but the mix is shifting toward larger vehicles and higher-specification braking as the middle class expands and highway infrastructure improves.

Local Indian brake companies like Brakes India (part of the TSF Group), Sundaram Brake Linings, Rane Brake Linings, ASK Automotive, and Masu Brakes have strong positions, while Korean supplier Mando maintains significant Indian operations. Global suppliers like ZF, AUMOVIO, and Bosch have established Indian operations, often through partnerships with local companies who provide manufacturing infrastructure and customer relationships.

The engineering talent pool in India is large and increasingly sophisticated. Many global brake suppliers have established engineering centers in India—not just for cost arbitrage, but for genuine development work. Simulation and analysis capabilities in particular have grown rapidly. For careers, India-based engineering roles are becoming more substantive than the "support center" model of a decade ago.

JAPAN AND SOUTH KOREA: THE VERTICAL ECOSYSTEM

Japan and South Korea deserve attention because their automotive supply chain model differs fundamentally from the Western approach—and brake suppliers are embedded in that model.

Japanese automakers (Toyota, Honda, Nissan) have traditionally operated through keiretsu-style supplier networks—vertically integrated ecosystems where long-term relationships, partial equity ownership, and deep technical collaboration define the customer-supplier dynamic. Akebono, ADVICS (an Aisin subsidiary within the Toyota Group), and Astemo (which absorbed Honda-affiliated Nissin Kogyo in 2021) operate within these networks, making it exceptionally difficult for outside suppliers to win programs at Japanese OEMs.

Korean automakers (Hyundai-Kia Group) have their own ecosystem, anchored by Mando (now HL Mando) as the primary brake supplier and Hyundai Mobis as the group's integrated module and brake system supplier. The relationships are deeply integrated, with both companies developing brake systems in close coordination with Hyundai-Kia's engineering teams.

What this means practically: if you work for a Western brake supplier, Japanese and Korean OE programs are among the hardest to win. The relationships are deep, switching costs are high, and the procurement culture favors continuity. This is slowly changing—Hyundai-Kia in particular has diversified its supply base more than Japanese OEMs—but it remains a structural reality.

WHAT GLOBAL DYNAMICS MEAN FOR YOUR CAREER

A few practical implications worth highlighting:

Willingness to work across borders matters. At any major brake supplier, career advancement increasingly requires global exposure—whether that's a rotational assignment overseas, managing a cross-regional team, or supporting a customer's global platform. Professionals who are effective only in their home market have a lower ceiling.

The center of gravity is shifting east. More engineering headcount, more manufacturing investment, and more new vehicle programs are in Asia than anywhere else. Career opportunities reflect that shift.

Regulatory knowledge is becoming a differentiator. Understanding how Euro 7 affects material development, how ECE R90 shapes aftermarket strategy, or how Chinese certification processes work gives you practical value that transcends geography.

Language skills are useful but not essential. English is the working language in most multinational brake companies. But conversational ability in Mandarin, German, Japanese, or Korean—even at a basic level—signals cultural awareness and opens informal communication channels that English-only professionals miss.

KEY TAKEAWAYS

- China is the largest vehicle market by volume and leads in EV brake technology adoption, with local suppliers now competing credibly against multinationals.

- Europe drives brake technology innovation through regulation—Euro 7 particle emissions and ECE R90 aftermarket homologation create dynamics that don't exist elsewhere.
- North America's shift toward light trucks and SUVs shapes brake system requirements; its open aftermarket structure creates both opportunity and quality variation.
- India is a significant growth market with an increasingly capable engineering talent pool being leveraged by global suppliers.
- Japanese and Korean supply chains operate on relationship-driven models that create high barriers for outside suppliers.
- Global exposure and cross-regional effectiveness are increasingly important for career advancement in the brake industry.

PART II
HOW THE INDUSTRY WORKS

6.
HOW OE BUSINESS REALLY WORKS

For outsiders—and even many industry veterans who've worked only in aftermarket or manufacturing—the OE (Original Equipment) business can seem opaque. How do brake suppliers win programs? Why are margins so thin? What does it actually take to be an OE supplier?

This chapter pulls back the curtain on how brake business really gets done with the automakers.

THE PROGRAM LIFECYCLE

Vehicle development follows a predictable cycle, typically spanning five to seven years from initial concept to end of production. Understanding this timeline is essential for anyone selling to or working with OEMs.

Concept and Architecture (Years 1-2)
Before a vehicle has a name or a design, platform engineers are making fundamental architecture decisions. Will it share a platform with other models? What powertrains will it support? What are the weight and performance targets?

Brake requirements begin to emerge at this stage—general parameters like axle loads, performance targets, and packaging constraints. Preferred suppliers may be consulted informally, but formal sourcing hasn't begun.

Advanced Development (Years 2–3)

Engineering teams refine the vehicle concept and develop detailed specifications. For brakes, this means defining stopping distance requirements, pedal feel targets, NVH expectations, and weight limits.

During this phase, OEMs often issue Requests for Information (RFIs) to potential suppliers. RFIs are exploratory—the automaker is gathering information about available technologies, supplier capabilities, and rough cost ranges. No business is being awarded; they're building knowledge to inform sourcing decisions.

Smart suppliers use RFI responses strategically: demonstrating technical capability, highlighting relevant experience, and beginning to build relationships with the engineering team. The relationships formed during RFI discussions often influence who gets invited to quote.

Sourcing (Years 3–4)

The formal sourcing process begins with a Request for Quotation (RFQ). Unlike an RFI, an RFQ is a formal bid solicitation with detailed specifications and commercial requirements.

A brake system RFQ package typically includes:

Technical specifications—stopping distance requirements, pedal feel parameters, NVH targets, weight limits, durability requirements, and interface dimensions.

Volume projections—expected annual volumes, launch timing, and program duration.

Commercial terms—target pricing expectations, payment terms, warranty requirements, and sometimes explicit cost reduction expectations.

Quality requirements—PPAP expectations, testing requirements, and quality system certifications needed.

Responding to an RFQ is a major undertaking. Suppliers must develop preliminary designs, estimate tooling and manufacturing costs, assess capacity requirements, and price the business—all while competitors are doing the same. The quotation process often takes months.

Quotes are evaluated across multiple dimensions: piece price (the per-unit cost), tooling costs, engineering cost recovery, technical capability, quality history, and strategic fit. The lowest price doesn't always win. OEMs consider total value—a supplier with higher piece price but better technical support, superior quality, or strategic alignment may be selected.

Sourcing decisions are typically made by cross-functional teams including purchasing, engineering, and quality. Large programs may require executive approval. Politics and relationships matter—the procurement team may prefer one supplier while engineering advocates for another.

Program Execution (Years 4-6)

Once sourced, the real work begins. Suppliers must complete detailed design, build and validate prototypes, develop manufacturing processes, and ramp up to production volume—all while meeting aggressive timing milestones.

The APQP (Advanced Product Quality Planning) process structures this work. APQP defines the phases of product development and the documentation required at each stage. Key milestones include:

Design reviews where engineering concepts are validated against requirements.

DVP&R (Design Verification Plan and Report) execution, demonstrating through testing that the design meets specifications.

Process development, defining how the product will be manufactured.

PPAP (Production Part Approval Process) submission, providing documented evidence that the production process can consistently produce parts meeting specifications.

Missing milestones has consequences. OEMs track supplier performance rigorously. Repeated delays damage the relationship, reduce chances of future business, and in extreme cases result in the program being re-sourced to another supplier—a devastating outcome after investing in tooling and development.

Production and Continuous Improvement (Years 5-10+)
Reaching "Job 1" (the start of production) is a major milestone but not the end of the story. OE suppliers are expected to support the program throughout its production life, which may span five years or more.

FIGURE 4: OE Program Lifecycle

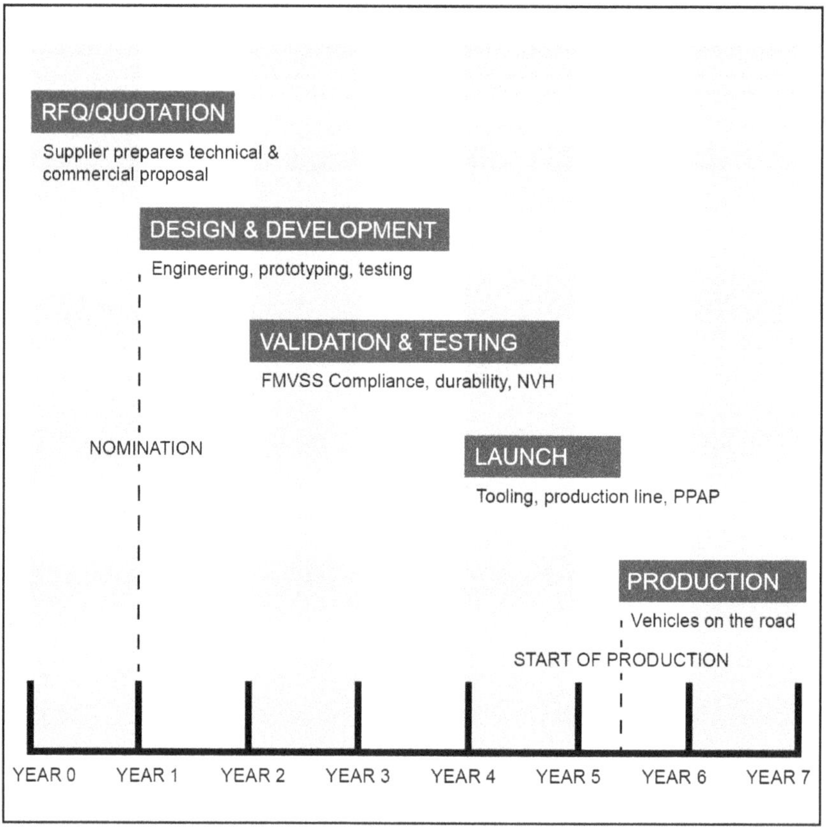

Support includes ongoing quality management, warranty support, engineering changes as needed, and—critically—continuous cost reduction. Most OE contracts include annual productivity expectations, typically a few percent price reduction per year. Suppliers must find efficiencies to maintain margins while passing savings to the customer.

Engineering changes after Job 1 are expensive and difficult. A change to a production component may require new tooling, updated validation testing, and PPAP re-submission. Suppliers who design robust, change-resistant products from the start are rewarded; those who require frequent running changes suffer both direct costs and relationship damage.

THE ECONOMICS OF OE SUPPLY

OE brake supply is a low-margin, high-volume business. Understanding the economics explains why the competitive landscape looks the way it does.

The Margin Reality

Operating margins on OE brake business typically run in the low single digits. This compares unfavorably to aftermarket margins, which are typically several times higher. Why do suppliers compete for low-margin business?

Volume and scale matter. A single OE program might represent millions of units over its production life. Even at thin margins, that volume justifies investment in automation and creates manufacturing efficiency that benefits other business.

The OE-to-aftermarket halo effect varies dramatically by company. Akebono actively leverages its OE credibility in aftermarket marketing. Others have historically deprioritized the connection, treating aftermarket as a distinct commercial channel rather than a downstream beneficiary of OE wins. ZF and Continental (now AUMOVIO) took it further structurally, keeping OE and aftermarket operations so segmented that the aftermarket teams had little visibility into what the OE side was doing. Brembo falls somewhere in between. The connection only works as a competitive advantage if the company actively pushes it.

Aftermarket pull is significant. Being specified as OE builds brand credibility and creates future aftermarket demand as vehicles age into the replacement cycle.

Relationship value accumulates. Performing well on one program positions suppliers for future business. OEMs prefer working with known suppliers when possible.

Price Structure Complexity

OE pricing involves more than a simple per-unit number:

Piece price is the per-unit cost paid for each component delivered. This is what everyone focuses on, but it's not the whole picture.

Tooling costs cover the molds, dies, and fixtures needed to produce the parts. These might be paid separately upfront, amortized into piece price, or some combination. Who owns the tooling—supplier or OEM—affects pricing flexibility and risk.

Engineering cost recovery compensates the supplier for development work. OEMs vary in their willingness to pay for engineering; some expect suppliers to absorb development costs as the price of competing for the business.

Annual price reductions are built into most contracts. A supplier quoting $15 per unit in year one might need to deliver at $14.25 by year three. This productivity expectation must be factored into initial pricing.

Managing Commodity and Currency Risk

Brake components depend on raw materials—steel, aluminum, iron, friction material ingredients—whose prices fluctuate. Most OE contracts include some mechanism for addressing commodity volatility:

Pass-through arrangements index pricing to raw material costs, adjusting automatically as commodities move.

Periodic resourcing reopens pricing at defined intervals, allowing adjustment for changed conditions.

Fixed pricing places commodity risk entirely on the supplier—acceptable when materials are stable, potentially devastating when they're not.

Currency adds another layer of complexity for global suppliers. A European supplier selling to a North American OEM in dollars faces exchange rate risk. Hedging strategies, natural hedges through localized production, and contract terms all play roles in managing this exposure.

CUSTOMER SCORECARDS AND PERFORMANCE MANAGEMENT

OEMs measure supplier performance systematically, and those measurements directly affect future business opportunities.

Quality Metrics

PPM (Parts Per Million) defective is the universal quality metric. An expectation of 10 PPM means no more than 10 defective parts per million delivered. Leading suppliers achieve single-digit PPM; anything above 50 PPM signals significant problems.

But PPM alone doesn't tell the whole story. A supplier with good PPM but frequent containment actions (sorting, inspections, rework) is worse than one with marginally higher PPM and stable processes. OEMs track not just defect rates but the cost and disruption of quality issues.

Warranty performance matters increasingly as OEMs push responsibility for field failures back to suppliers. A brake pad that wears prematurely or a caliper that leaks after 30,000 miles creates warranty costs—and increasingly, those costs are charged back to suppliers.

Delivery Performance

On-time delivery expectations are stringent—typically 98%+ or better. Modern just-in-time manufacturing systems have minimal buffer inventory; a missed delivery can shut down an assembly line.

Line stoppages due to supplier delivery failures are catastrophic events. The costs—idle workers, disrupted production, expedited shipping to recover—can run hundreds of thousands of dollars per hour. Suppliers responsible for line stoppages face serious consequences including chargebacks, business holds, and loss of future opportunities.

Scorecard Politics

Every major OEM maintains supplier scorecards summarizing quality, delivery, cost, and technology performance. These scorecards directly influence sourcing decisions—suppliers in the top tier get invited to quote new business; those in lower tiers may be excluded regardless of their technical capabilities or pricing.

Improving a poor scorecard takes time and sustained effort. Quick fixes don't exist; OEMs look for consistent performance over multiple quarters. Suppliers recovering from poor performance often find themselves in probationary status, subject to additional monitoring and restrictions.

WINNING BUSINESS: BEYOND THE QUOTE

The formal RFQ process matters, but it's not the whole story. Business is also won through:

Relationship Building

Engineers who develop solutions alongside OEM counterparts during advanced development naturally position their companies for sourcing success. Trust built over months of collaboration influences decisions that are nominally based on commercial evaluation.

Executive relationships matter too. When OEM leadership knows and trusts supplier leadership, intangible factors like "strategic fit" and "partnership potential" carry more weight.

Technology Demonstration

Suppliers who demonstrate innovation—showing capabilities beyond current requirements—position themselves as preferred partners for future programs. Technology roadmap presentations, prototype development for emerging applications, and participation in industry consortiums all contribute.

Resident Engineering

Placing engineers at or near OEM locations provides immediate access to customer needs and concerns. Resident engineers troubleshoot problems quickly, participate in design reviews, and build relationships that inform future opportunities. The investment is substantial but often pays dividends.

Account Management

Dedicated account management—someone whose job is understanding and serving a specific OEM—outperforms generalized sales approaches. Effective account managers know the customer's programs, priorities, politics, and pain points. They anticipate needs rather than just responding to inquiries.

KEY TAKEAWAYS

- Vehicle programs span 5–7 years from concept to end of production—brake suppliers engage during advanced development, 3–4 years before Job 1.
- RFQs are won through engineering capability, quality consistency, cost competitiveness, and relationships—lowest price doesn't always win.
- OE margins are thin (typically low single digits) with annual price reduction expectations built into contracts—suppliers must continuously improve productivity.
- Customer scorecards measuring quality (PPM), delivery (98%+ expected), and other metrics directly influence future business opportunities.

7.
NAVIGATING THE QUALITY ECOSYSTEM

If you've never worked at an automotive supplier, your first encounter with the quality system will be disorienting. You'll walk into meetings where everyone speaks in acronyms—APQP, PPAP, DFMEA, PFMEA, SPC, MSA, Cpk—and assumes you know what they mean. You'll discover that changing a dimension on a drawing requires customer approval, weeks of paperwork, and sometimes a formal audit. You'll learn that a single quality escape—one bad part reaching the assembly line—can cost your company millions and get your plant placed on "new business hold."

This chapter won't teach you how to run a quality system. That's what your company's quality manual is for, and it would take a book of its own. Instead, this is the orientation that a good mentor would give you in your first week: what the system is, why it exists, and how it shapes daily life at a brake supplier.

THE GOVERNING FRAMEWORK: IATF 16949

Every serious automotive supplier operates under IATF 16949, the quality management standard specific to the automotive industry. It's built on ISO 9001 but adds automotive-specific requirements that make it considerably more demanding. Your facility is audited against this standard regularly, and losing certification effectively means losing the ability to supply OE customers.

What IATF 16949 means practically is that everything is documented, everything is controlled, and everything is traceable. The raw material in a brake pad can be traced back to its supplier, its lot, its incoming inspection results. The machine that pressed that pad, the operator who ran the press, the inspection results, the shipping records—all documented. When a quality issue surfaces, the system is designed to let you trace backward from the problem to its root cause and forward from a suspect lot to every part that might be affected.

This level of documentation and control can feel bureaucratic, especially to people coming from less regulated industries. But it exists for a reason. Brake components are safety-critical. A failure can kill people. The system is designed to prevent that.

APQP: HOW PRODUCTS GET DEVELOPED

APQP—Advanced Product Quality Planning—is the structured process that governs how a new product moves from concept to production. If Chapter 6 described the business side of winning an OE program, APQP is the engineering and quality framework that makes it happen.

APQP unfolds in five phases, and understanding them helps you see where your work fits in the bigger picture.

Phase 1: Plan and Define. The customer's requirements are translated into technical specifications. What are the performance targets? What are the dimensional requirements? What standards apply? This is where the program's quality expectations are established.

Phase 2: Product Design and Development. Engineering designs the brake component, runs simulations, builds prototypes, and tests them. The critical output at this stage is the Design FMEA—a systematic analysis of everything that could go wrong with the design and how those risks are mitigated. More on FMEA in a moment.

Phase 3: Process Design and Development. This is where manufacturing engineering determines how the product will actually be made. What equipment? What process parameters? What controls? The Process FMEA is developed here, analyzing manufacturing risks the same way the Design FMEA analyzed design risks. The control plan—the document that specifies every inspection, measurement, and test performed during production—is drafted.

Phase 4: Product and Process Validation. The production line runs a significant trial production. Parts are measured, tested, and verified against every specification. This is where you demonstrate that the process can consistently produce good parts at production volume and speed. Statistical capability studies (Cpk analysis) prove that your process is not just producing acceptable parts, but producing them with enough margin that you won't drift into trouble.

Phase 5: Launch and Continuous Improvement. Production begins, lessons learned are documented, and the focus shifts to sustaining quality and reducing variation over time.

The timeline for this process in brake components is typically 18 to 36 months, depending on complexity. It's not unusual for a friction material development program to run longer, given the extensive validation required.

PPAP: PROVING YOU CAN DO WHAT YOU SAID

PPAP—Production Part Approval Process—is the formal submission that proves your product and process meet customer requirements. Think of it as the exam at the end of APQP. Until your customer approves your PPAP, you cannot ship production parts.

A full PPAP submission includes 18 elements, and understanding the major ones helps demystify the process:

Design records (your engineering drawings), engineering change documents (if the design evolved), dimensional results (measurements of actual production parts against every specification), material test results (confirming material properties meet requirements), performance test results (proving the part functions correctly), process flow diagrams (showing how the part is manufactured step by step), the control plan (documenting every quality check in the production process), and measurement system analysis (proving your gauges and measurement methods are reliable enough to detect the variation you're trying to control).

The customer's SQE—Supplier Quality Engineer—reviews the PPAP and either approves it, approves it with conditions, or rejects it. This is why the SQE has so much influence: they are the gatekeeper between your development work and actual production revenue. A good relationship with your customer's SQE is one of the most valuable assets a supplier can have.

FMEA: STARTING WITH WHAT COULD GO WRONG

Failure Mode and Effects Analysis—FMEA—is the risk assessment methodology that runs through everything. There are two types, and both are central to brake component development.

Design FMEA (DFMEA) asks: what could go wrong with the design itself? For a brake rotor, this might include: venting channels could be undersized, leading to inadequate heat dissipation. The hat section geometry could allow excessive lateral runout. The metallurgy could permit thermal cracking under repeated hard stops. For each potential failure mode, the team assesses severity (how bad is it if it happens?), occurrence (how likely is it?), and detection (how likely are we to catch it before it reaches the customer?). These three scores multiply into a Risk Priority Number that drives action.

Process FMEA (PFMEA) asks the same questions about manufacturing. Could the press temperature drift out of range, causing incomplete resin cure in a brake pad? Could a machining tool wear gradually enough that parts drift out of specification between inspections? Could raw materials from different lots introduce variation that affects friction performance?

In brake components, the severity scores are inherently high because failure affects vehicle safety. That means occurrence and detection carry the burden of bringing risk to acceptable levels—which is why process controls in brake manufacturing are so extensive.

ENGINEERING CHANGES: WHY SIMPLE THINGS TAKE MONTHS

One of the most common frustrations for people new to the industry is how long it takes to change anything. You identify an improvement—a material substitution that reduces cost, a tolerance change that improves manufacturability, a design tweak that solves a field issue—and then discover that implementing it requires a formal engineering change process that can take months.

Here's why. Any change to a brake component that's been PPAP-approved must be re-validated. The customer approved a specific design, made with specific materials, using a specific process, with specific controls. Change any of those variables and you need to demonstrate—with data—that the change doesn't compromise performance or quality. Depending on the magnitude of the change, that can range from a partial PPAP resubmission (updated dimensional results and material test data) to a complete revalidation including dynamometer testing and vehicle-level validation.

Customers classify changes by risk level. A "cosmetic" change might require minimal documentation. A change to friction material formulation—which directly affects stopping performance—will require the most rigorous revalidation, potentially including months of testing.

Some customers require formal notification and approval before you even begin validation work.

This is not bureaucracy for its own sake. It reflects hard-won lessons from incidents where well-intentioned changes introduced unintended consequences. A raw material substitution that seemed equivalent turned out to behave differently at temperature extremes. A process parameter adjustment that improved one quality metric degraded another. The change control system exists to catch these issues before they reach vehicles.

WHEN THINGS GO WRONG: QUALITY ESCAPES

Despite all these controls, defective parts sometimes reach the customer. When they do, the response is swift and structured.

An 8D report (Eight Disciplines) is the standard problem-solving format. It requires the supplier to: assemble a team, describe the problem precisely, implement immediate containment actions (sorting suspect inventory, inspecting parts already shipped, reworking or scrapping affected material), identify the root cause (not the symptom, the actual cause), implement corrective actions, verify those actions work, prevent recurrence through systemic changes, and recognize the team.

The containment phase is the most urgent and often the most expensive. If suspect parts have already been shipped to an assembly plant, the supplier may need to send teams to sort parts at the customer's facility—sometimes around the clock. I've seen brake suppliers dispatch dozens of quality engineers and inspectors to customer plants within hours of a quality notification, at costs that can reach six figures per day.

The consequences extend beyond the immediate incident. Quality escapes affect your customer scorecard, which affects future business awards. Repeated issues can result in "controlled shipping"—where every part must be inspected by a third party before shipment, at the supplier's expense. In extreme cases, a customer can place a supplier on "new business hold," effectively freezing them out of future programs until systemic improvements are demonstrated.

This is why quality roles at brake suppliers carry so much weight. The quality team isn't just checking boxes—they're protecting the company's ability to win future business and, more fundamentally, protecting the people who depend on these products every time they press the brake pedal.

THE ALPHABET SOUP: A QUICK REFERENCE

For convenience, here are the acronyms you'll encounter most frequently:

AIAG: Automotive Industry Action Group—publishes the core quality tool manuals (APQP, PPAP, FMEA, SPC, MSA)

APQP: Advanced Product Quality Planning—the structured product development process

Cpk: Process Capability Index—a statistical measure of how well your process stays within specification limits. Customers typically require Cpk of 1.33 or higher for critical dimensions; 1.67 for safety-critical characteristics

DFMEA: Design Failure Mode and Effects Analysis

8D: Eight Disciplines—structured problem-solving methodology for quality issues

FMEA: Failure Mode and Effects Analysis

IATF 16949: The automotive quality management standard

MSA: Measurement System Analysis—proving your gauges and measurement methods are reliable

PFMEA: Process Failure Mode and Effects Analysis

PPAP: Production Part Approval Process—the formal submission proving your product and process meet requirements

RPN: Risk Priority Number—severity × occurrence × detection scoring in FMEA

SPC: Statistical Process Control—using statistics to monitor and control manufacturing processes

SQE: Supplier Quality Engineer—the customer's representative who evaluates and approves your quality performance

KEY TAKEAWAYS

- IATF 16949 is the governing quality standard for automotive suppliers—it requires comprehensive documentation, traceability, and process control that can surprise people from less regulated industries.
- APQP provides the structured framework for developing new products through five phases, from planning through launch. Understanding where your role fits in this process helps you see the bigger picture.
- PPAP is the formal proof that your product and process meet requirements—your customer's SQE must approve it before you can ship production parts.
- FMEA (both Design and Process) is the risk assessment methodology that drives how brake components are designed and manufactured. High severity scores for safety-critical parts mean that occurrence prevention and detection controls must be exceptionally strong.
- Engineering changes require formal revalidation because past experience has shown that seemingly minor changes can have unintended consequences in safety-critical components.
- Quality escapes trigger rapid, structured responses (8D) with significant financial and commercial consequences. Quality performance directly affects a supplier's ability to win future business.

8. THE MATERIAL WARS

If there's one area where the brake industry gets genuinely competitive—where companies guard their secrets and engineers speak in careful terms—it's friction materials.

The formulation of a brake pad is essentially a recipe, and like any recipe, small changes can produce dramatically different results. I've visited friction material labs where engineers showed me compounds containing dozens of different ingredients—each one carefully calibrated to affect performance in specific ways.

THE FRICTION MATERIAL LANDSCAPE

Semi-metallic pads were once the volume leader but have been largely displaced by NAO/ceramic formulations in both OE and aftermarket applications. Today, semi-metallic formulations are primarily found on heavier-duty platforms (2500-series trucks and above) where their heat handling capability remains essential. They contain a mix of metals—steel, iron, copper (in legacy formulations)—combined with friction modifiers and fillers. Semi-metallics offer a good balance of performance and cost, handling heat well and providing strong stopping power. The trade-off is more noise, more dust, and harder wear on rotors. They're the workhorse choice for mainstream vehicles and demanding applications.

Non-asbestos organic (NAO) pads—which originated in Asia in the early 1990s—use fibers, fillers, and binding agents with minimal metal content (historically copper, now transitioning to aluminum or zinc in copper-free formulations). They're quieter, produce less visible dust, and are gentler on rotors. NAO formulations are common in North America and increasingly in Asia, where NVH (noise, vibration, harshness) concerns drive customer preferences. The trade-off is typically lower heat resistance and faster wear under aggressive use.

Low-steel formulations blend the approaches—organic base materials with small amounts of steel for improved heat handling. They are the dominant formulation type in European applications, where sustained high-speed autobahn driving demands better thermal performance than pure NAO formulations.

In practice, the industry often groups ceramic formulations under the broader NAO/ceramic umbrella, as the distinction between premium NAO and ceramic has blurred. Ceramic pads use ceramic fibers combined with various fillers and bonding agents. They are quiet, produce minimal visible dust, handle heat well, and last longer than conventional materials. The downsides are higher cost and potentially less initial "bite" in cold conditions. Ceramic formulations are standard on luxury vehicles and gaining ground in mainstream applications as costs come down. The segment is projected to grow faster than other material types through the decade.

Carbon-ceramic composites sit at the extreme end—extraordinarily expensive, extraordinarily capable. These materials can withstand temperatures that would destroy conventional brakes, weigh significantly less, and last far longer. You find them on supercars (where they can add $15,000+ to the vehicle price) and in motorsports. Brembo, SGL Carbon, and others are investing heavily in expanding carbon-ceramic production capacity, suggesting growing demand as the premium vehicle segment expands.

THE COPPER QUESTION

One of the most significant shifts I've watched unfold over the past several years is the move away from copper in brake pads.

For decades, copper was a go-to ingredient in friction formulations. It conducts heat well, provides consistent friction characteristics, and helps with overall pad performance. The problem? Copper from brake dust washes into waterways and accumulates in ecosystems, harming aquatic life.

California and Washington led the regulatory charge, passing legislation that phases out copper in brake pads. As of 2025, brake pads sold in these states must contain less than 0.5% copper by weight. The rest of the country—and increasingly, the world—has followed suit.

This might sound like a minor technical adjustment. It isn't. Reformulating friction materials without copper while maintaining performance, noise characteristics, and wear properties has required enormous R&D investment across the industry. I've talked to engineers who spent years working on copper-free formulations, testing and iterating until they could match the performance of their legacy products.

The transition is largely complete now—Akebono, for example, has developed ultra-low copper brake pads certified for both EU and California markets, installed in over 14 million vehicles. But it's a perfect example of how environmental regulation drives innovation in this industry—and how changes that seem invisible to consumers require tremendous effort behind the scenes.

INSIDE THE FRICTION FORMULATION

A brake pad formulation is essentially a recipe—but one involving 15–25 or more ingredients, each contributing specific properties. Understanding these ingredients explains why formulation development is both science and art.

Each ingredient category serves a specific function: binders (typically phenolic resin) form the structural matrix; reinforcing fibers (steel, aramid, ceramic, mineral, glass) provide the skeleton; friction modifiers (graphite,

FIGURE 5: Brake Pad Friction Material Composition

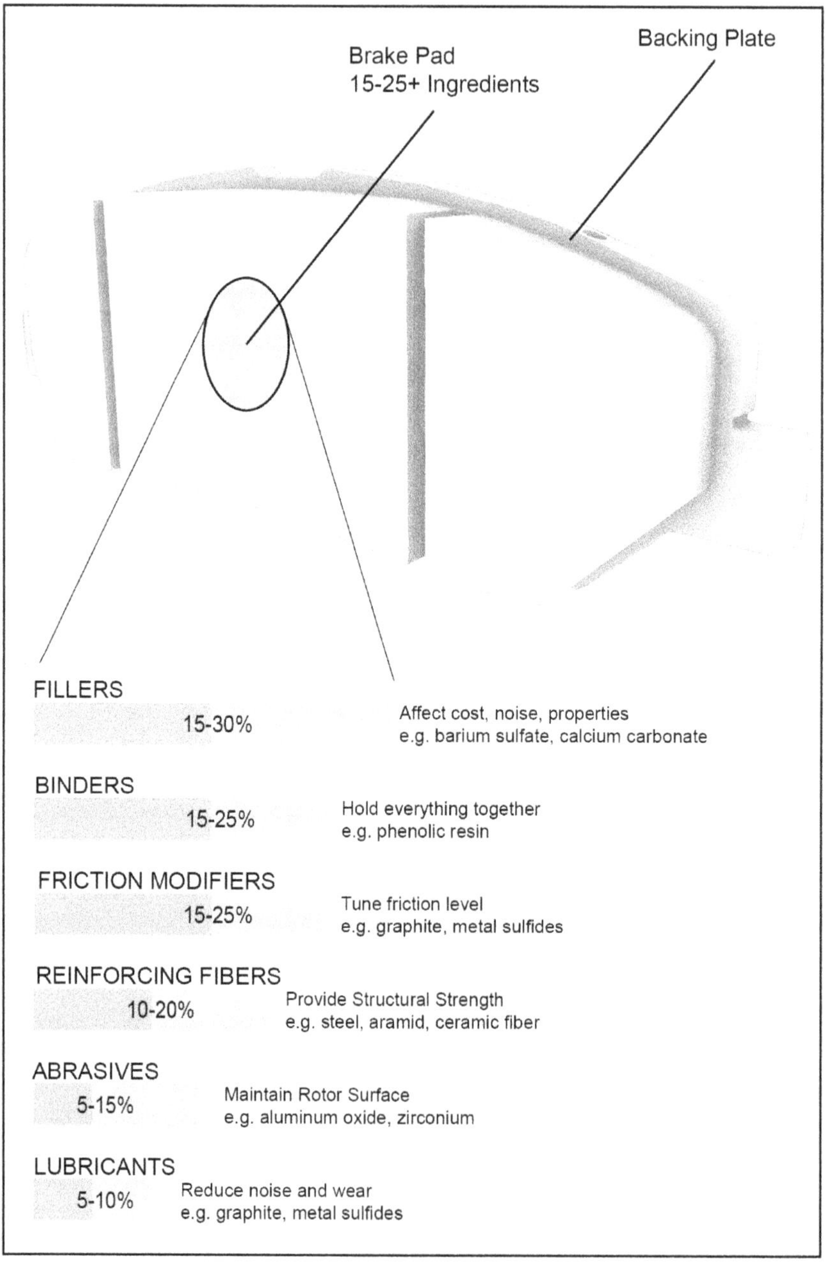

metal sulfides, aluminum oxide, silicon carbide, and cashew dust, to name a few) tune the grip characteristics at different temperatures and maintain a clean rotor interface; and fillers (barium sulfate, calcium carbonate, mica) add bulk while fine-tuning thermal and physical properties. The interplay between these categories is what makes formulation both an art and a science—and why experienced formulators are among the industry's most guarded assets. For the complete ingredient-by-ingredient technical breakdown, see Appendix E: Technical Deep Dives.

THE NOISE PROBLEM: NVH DEEP DIVE

Brake noise is one of the most frustrating challenges in friction material development. A pad that provides excellent stopping power but squeals annoyingly will generate customer complaints and warranty claims.

Why Brakes Squeal

Brake noise is a complex system-level resonance phenomenon that has been the subject of countless PhD theses and remains incompletely understood. There are many hypotheses about what causes brake noise, and all have some validity. Rather than attempting to definitively explain it, what matters for industry professionals is understanding the solutions that minimize this complex phenomenon. At the most basic level, brake squeal involves stick-slip vibration that couples with structural resonances to produce audible noise.

At the microscopic level, the pad-rotor interface alternates between sticking and slipping. This happens because static friction (the force needed to start motion) exceeds kinetic friction (the force during motion). The pad briefly sticks, stress builds, the pad slips, stress releases, and the cycle repeats—thousands of times per second.

This stick-slip excitation couples with the natural vibration modes of the brake assembly—the rotor, caliper, pads, and mounting hardware all have frequencies at which they naturally vibrate. When stick-slip excitation occurs near one of these natural frequencies, the system resonates and noise becomes audible.

Squeal frequencies typically fall between 1,000 and 16,000 Hz—the range where human hearing is most sensitive. Lower-frequency vibrations manifest as "groan" or "moan"; higher frequencies become inaudible but may still cause structural concerns.

Engineers describe brake squeal as a mode coupling instability—when two vibration modes of the brake assembly converge at similar frequencies and lock together, feeding energy into each other and producing sustained noise. The coupling depends on friction coefficient, contact pressure, temperature, geometry, and material properties, which is why small changes in any variable can shift a system from quiet to noisy. For the full technical explanation of mode coupling dynamics, see Appendix E: Technical Deep Dives.

Solutions and Mitigations

Since squeal involves system resonance, solutions focus on either preventing excitation or damping resonant response:

Chamfers and slots cut into pad edges change the dynamic interaction between pad and disc and alter the natural frequency of the pad assembly, which can be helpful in mitigating noise. A chamfered leading edge, for example, changes how the pad first contacts the rotor, altering the excitation dynamics.

Shims—thin layers of material bonded to the pad backing—provide constrained layer damping. When the backing plate vibrates, it must shear the shim material, which dissipates energy as heat rather than radiating it as sound. The construction of a shim can be tailored to address specific noise frequencies. The rubber chemistry used in shims can dampen certain frequencies or perform differently at specific temperatures. Multi-layer shim construction can damp vibrations across varying frequency ranges, and even the surface texture of the shim has been proven effective in mitigating noise. There is no single shim design that solves all brake noise—each application requires its own solution.

Friction material formulation plays a significant role in brake noise, though the relationship is more complex than many assume. A common belief is that higher friction coefficients produce more noise, but that is not necessarily the case. Many attributes within a formulation affect noise performance—damping capacity, porosity, the specific ingredient blend, and the material's response to environmental conditions like temperature and humidity. Changing the friction coefficient, modifying temperature behavior, or adjusting material damping all affect squeal propensity.

Rotor design modifications change resonant frequencies. Slots, drilled holes, and asymmetric vane patterns (where vanes have varying spacing) all disrupt the regular patterns that enable resonance.

Hardware design matters too. The caliper mounting, abutment clips, and anti-rattle hardware all influence system dynamics. Worn or missing hardware often causes noise complaints.

Why NVH Is So Difficult

Brake noise is frustrating because it's intermittent, condition-dependent, and often impossible to reproduce on demand:

Temperature sensitivity: Many brake systems squeal only in specific temperature ranges—cold morning operation, or after cooling from heavy use.

Humidity effects: Moisture absorbed overnight can completely change friction behavior until brakes warm up and dry out.

Wear state: New pads often squeal until bedded in. Worn pads may squeal as the friction material thins.

Manufacturing variation: Small differences in friction material batch-to-batch can tip systems from quiet to noisy.

Vehicle variation: Two identical vehicles may behave differently due to manufacturing tolerances in suspension components, hub faces, or other factors.

This variability makes NVH development expensive and time-consuming. Testing must cover many conditions, and solutions must be robust across variation.

ROTOR MATERIALS: BEYOND BASIC IRON

While gray cast iron dominates the rotor market, variations in iron composition and alternative materials serve specific applications.

The gray iron used for brake rotors is specifically formulated for the application, with carbon content, graphite flake structure, and trace element composition all carefully controlled. These metallurgical details determine the rotor's thermal conductivity, damping characteristics, wear behavior, and noise tendencies. For the complete iron metallurgy breakdown, see Appendix E: Technical Deep Dives.

High-Carbon Rotors

"High-carbon" has become a marketing term in the aftermarket, but the underlying concept has merit. Higher carbon content can increase graphite volume, improving thermal properties and noise damping.

However, simply adding more carbon isn't straightforward. Excess carbon can form carbides rather than graphite, making the iron harder and more brittle. Higher-carbon iron is also more difficult to pour and slightly more challenging to machine, adding manufacturing cost. Achieving desired graphite structure with higher carbon content requires careful process control.

Some high-carbon claims are more marketing than metallurgy. Consumers should be skeptical of dramatic performance claims based solely on carbon content.

Coated Rotors

Various coatings address specific rotor challenges:

Zinc coatings (zinc flake or zinc-aluminum) protect non-friction surfaces from corrosion. The hat area and vane areas don't contact the pad; without protection, they rust visibly—cosmetically unacceptable even if functionally irrelevant.

Geomet® and similar coatings provide corrosion protection with a more uniform appearance than zinc plating.

Edge coatings seal the outer edge and inner bore against corrosion.

Full-face coatings extend protection to the friction surface, burning off during initial use. These mainly provide cosmetic protection during shipping and storage.

Harder coatings like nitriding (FNC—ferritic nitrocarburizing) actually change surface properties, creating a harder, more corrosion-resistant layer. These can reduce initial wear and provide longer-lasting protection, though at higher cost.

Carbon-Ceramic Composites

Carbon-ceramic rotors represent the extreme end of brake rotor technology—extraordinarily capable, extraordinarily expensive.

The manufacturing process involves creating a carbon fiber preform, infiltrating it with silicon at high temperature, and allowing the silicon to react with carbon to form silicon carbide. The result is a ceramic matrix composite with exceptional properties:

Temperature capability far exceeding iron—carbon-ceramic rotors can operate at temperatures that would melt conventional rotors.

Significant weight reduction compared to iron rotors—important for unsprung mass reduction.

Exceptional longevity—carbon-ceramic rotors can last the life of the vehicle under normal use.

Fade resistance that makes them favorites in motorsports and high-performance applications.

The drawbacks are significant:

Cost runs from several thousand to tens of thousands of dollars for a complete set of rotors, versus hundreds of dollars for iron.

Cold performance and initial bite can be reduced until rotors reach operating temperature.

Matching pads must be specifically designed for carbon-ceramic; conventional pads don't work.

Damage from impacts or contamination can be costly to repair.

Carbon-ceramic remains a niche product for supercars and racing. As manufacturing scales and costs decline, applications may expand—but iron rotors will dominate for the foreseeable future.

THE PEOPLE BEHIND THE FORMULATIONS

Behind every friction formulation is a specialist known as a formulator or compounder—one of the brake industry's most critical and least understood roles. These professionals develop the "recipes" that determine how a brake pad performs, balancing dozens of variables across ingredients, processing parameters, and performance requirements.

Formulators work at the intersection of chemistry, materials science, and practical engineering. They understand not just what each ingredient does in isolation, but how ingredients interact under heat, pressure, and repeated cycling. Much of this knowledge is experiential—learned through years of trial, observation, and refinement rather than from textbooks.

The scarcity of experienced formulators is a persistent industry challenge. It takes a decade or more to develop the intuition that separates competent formulators from exceptional ones. Companies guard these professionals carefully, and recruiting them is among the most difficult tasks in brake industry talent acquisition. For those considering careers in this field, Chapter 19 explores the formulator role in depth.

KEY TAKEAWAYS

- Friction material formulations contain 15–25+ ingredients, each contributing specific properties—the recipe is a closely guarded competitive advantage.
- Semi-metallic pads remain the workhorse for heavier-duty and demanding applications due to their heat handling; ceramic pads are the premium tier offering quiet operation and low dust.
- The copper phase-out (below 0.5% by 2025 in California/Washington) required enormous R&D investment to maintain performance with copper-free formulations.
- Carbon-ceramic rotors offer extraordinary performance but cost several thousand to tens of thousands of dollars per set, limiting them to supercars and motorsports.

- Experienced friction material formulators are among the industry's scarcest talent—it takes 10–15 years to develop the intuition that separates competent formulators from exceptional ones.

The material choices covered in this chapter play out very differently depending on which side of the business you're on. In OE, the automaker specifies the formulation during vehicle development. In the aftermarket, those choices multiply into dozens of competing options at every quality tier. Chapter 9 explains how that aftermarket world actually operates.

9.
HOW THE AFTERMARKET REALLY WORKS

If Chapter 6 explains the OE world—long development cycles, thin margins, engineering-intensive relationships—the aftermarket is its opposite in almost every way. It's faster, more fragmented, higher-margin, and fundamentally different in who makes the purchasing decision—though margins face increasing pressure from big-box retailers who exert significant downward pricing force, sometimes with extended payment terms that can stretch to 12 months or longer.

Understanding how the aftermarket operates is essential because it represents a smaller share of brake industry revenue but a disproportionate share of profitability. Many brake companies derive the majority of their profits from aftermarket business, even when OE represents the majority of their volume. And the two business models interact in ways that aren't immediately obvious.

THE FUNDAMENTAL DIFFERENCE: WHO DECIDES

In OE, the automaker specifies the brake components during vehicle development. The end consumer has no say. The supplier wins business years before a single vehicle is sold, through engineering capability and commercial negotiation.

In the aftermarket, nobody specifies anything. When a vehicle needs replacement brake pads or rotors, someone—an installer, a parts counter person, a vehicle owner—chooses from dozens of available options. That choice happens in real time, influenced by brand recognition, availability, price, past experience, and the recommendation of whoever is closest to the transaction.

This distinction changes everything about how the business operates. OE is about winning programs through engineering and relationships with a handful of automakers. Aftermarket is about winning millions of individual purchase decisions through distribution reach, brand strength, application coverage, and availability.

HOW PARTS REACH THE VEHICLE

The traditional aftermarket distribution chain has multiple levels, each adding value and margin:

Manufacturers produce brake pads, rotors, calipers, hardware, and other components. Some are the same companies that supply OE—Brembo, Akebono, Bosch, AUMOVIO's ATE brand—selling aftermarket parts that resemble OE products. These companies reverse-engineer OE designs just like other traditional aftermarket manufacturers—they are not selling actual OE-produced parts. (Notably, even some premium brands have deep manufacturing ties in China—Brembo and Gold Phoenix, for example, formed a 50/50 joint venture, Shandong BRGP Friction Technology, dedicated to aftermarket pad production). Others are aftermarket specialists who manufacture replacement parts without any OE business. Still others are importers sourcing from overseas factories in China, India, or other low-cost regions under their own brand names.

Warehouse Distributors (WDs) are the logistics backbone of the aftermarket. Companies like Genuine Parts Company (NAPA), AutoZone, O'Reilly, and Advance Auto Parts on the retail side, and traditional WDs like Auto-Wares, Federated, Pronto, and The Parts Authority on the professional side, purchase from manufacturers in volume and stock parts in regional warehouses. Their value proposition is simple but critical: they make same-day or next-day delivery possible for virtually any brake application. A jobber or repair shop can't stock every part number for every vehicle; the WD solves that problem through scale and logistics infrastructure.

Jobbers and Parts Stores are the local outlets serving repair shops and retail customers. They may be independent businesses, members of buying groups, or corporate-owned stores in a national chain. Jobbers typically source from multiple WDs and manufacturers to ensure broad coverage and competitive pricing. The jobber's counter person is often the most influential person in the purchase decision—when an installer calls asking for pads for a 2019 Honda CR-V, whoever answers the phone recommends a brand.

Repair Shops are the primary installers of aftermarket brake components. Independent garages, dealership service departments, specialty brake shops, and quick-service chains like Midas or Meineke all purchase and install replacement brake parts. Some buy from jobbers; larger operations may purchase directly from WDs or even manufacturers for their highest-volume applications.

Retail and DIY represents consumers purchasing parts for self-installation. This segment is served by retail auto parts chains and increasingly by e-commerce platforms like Amazon, RockAuto, and specialty online retailers. While brakes are more complex than many DIY repairs, a meaningful percentage of replacement brake work is done by vehicle owners—particularly pad and rotor replacement.

BUYING GROUPS AND CONSOLIDATION

One of the most significant structural features of the aftermarket is the buying group phenomenon. Independent WDs and jobbers band together into cooperative purchasing organizations—groups like Aftermarket Auto Parts Alliance (AAPA), Automotive Parts Services Group (The Group), and others—to aggregate their purchasing volume and negotiate manufacturer pricing that individual members couldn't achieve alone.

Buying groups wield enormous influence. Their decisions about which brands to stock, which manufacturers to partner with, and which programs to promote can make or break an aftermarket brake brand. A manufacturer gaining preferred status with a major buying group gains access to thousands of outlets; losing that status can be devastating.

The aftermarket has also consolidated significantly. National chains have acquired independent distributors. Private equity has invested heavily in distribution. The result is a market where a handful of large players control an increasing share of distribution, while thousands of independent operators still serve local markets. This consolidation gives large distributors increasing pricing leverage with manufacturers.

THE CATALOG AND DATA INFRASTRUCTURE

Unlike OE, where parts are specified by engineering drawing numbers, the aftermarket runs on catalog data. When an installer needs brake pads for a specific vehicle, they look up the application in a catalog—physical or digital—to find which part numbers fit.

This data infrastructure is more complex and more important than outsiders realize:

FMSI D-Numbers are the aftermarket's common language for friction products. The Friction Materials Standards Institute assigns a standardized D-number to each unique brake pad or shoe application. When multiple manufacturers make pads for the same application, they all reference the same D-number. This standardization allows catalogs, counterperson lookup systems, and e-commerce platforms to match parts to vehicles regardless of manufacturer. FMSI has since added Z-number designations to identify applications where the OEM specifies mechanical attachment of the friction material to the backing plate, giving the aftermarket a way to signal OE-spec construction.

ACES (Aftermarket Catalog Exchange Standard) and **PIES (Product Information Exchange Standard)** are the data formats that power electronic cataloging. ACES defines how application data is structured (which parts fit which vehicles), while PIES defines product attributes (dimensions, features, images, marketing descriptions). Manufacturers who maintain clean, accurate ACES and PIES data get their products listed correctly in distributor systems; those with poor data find their products invisible to buyers.

Coverage is a critical competitive dimension. An aftermarket brake manufacturer's catalog might cover 85%, 92%, or 98% of vehicles in operation. Those percentage points matter enormously—an installer who can't find the right pad for a customer's vehicle switches to a brand that has it, and may not switch back. Coverage gaps are one of the most common reasons aftermarket brands lose business.

QUALITY TIERS AND WHAT THEY MEAN

The aftermarket spans a wide quality range, and understanding the tiers is essential for anyone in the business:

OE/OES (Original Equipment Service) parts are sold through dealer networks and are the same specifications used on new vehicles, typically from the same manufacturers that supply the automakers. They carry premium pricing but guarantee exact fit, function, and warranty compatibility. The dealer channel captures this premium by offering customers the confidence of "genuine" parts.

The aftermarket generally follows a "Good, Better, Best" product strategy. At the top sits the OE-equivalent or premium tier. An increasingly important category is what the industry calls "OE 2nd Line" or "OE

Value Line"—these are aftermarket pads that have been tested by the OE to their standards and are offered at dealerships at a lower price point than the genuine OE parts that came with the vehicle. This allows dealers to compete with independent service centers while keeping customers in the service bay.

Premium aftermarket parts claim to meet or exceed OE specifications and often come from recognized brand-name manufacturers. Some offer enhancements over base OE—reduced dust, lower noise, improved feel. Premium brands justify higher pricing through engineering, testing, brand reputation, and warranty coverage. It's worth noting, however, that most aftermarket companies have zero access to actual OE specifications. Their "meets or exceeds OE" claims are typically based on benchmarking the OE friction material for noise, durability, and performance using their own testing protocols—not on meeting the OE's actual validation requirements. It is more marketing perception than verified fact, though that doesn't mean the products are inferior.

Standard aftermarket parts meet basic functional requirements at competitive price points. Quality can be perfectly adequate, but varies significantly by manufacturer. This is the largest segment by volume—professional installers who want reliable parts without premium pricing.

Economy and value parts prioritize low price above all else. They may meet minimum functional requirements but often sacrifice durability, noise characteristics, or consistent quality control. In the North American market, where there is no mandatory aftermarket homologation requirement for brake friction products, economy parts face no regulatory barrier to market entry.

The quality tier landscape creates a persistent tension in the aftermarket. Price pressure pushes distributors and installers toward lower-cost options, while comebacks (warranty returns due to noise, premature wear, or poor performance) push them back toward quality. Every installer has stories about the "cheap pads that squealed on every car" or the "economy rotors that warped in 10,000 miles."

THE REGULATORY CONTRAST

One of the most striking differences between aftermarket brake markets globally is the regulatory environment:

North America has no mandatory aftermarket homologation requirement for replacement brake friction products. Any manufacturer can sell brake pads into the U.S. or Canadian market without demonstrating that they meet any performance standard. FMVSS 135 governs new vehicle brake performance, but there is no equivalent requirement for replacement parts. This open market creates opportunity for innovative manufacturers but also allows substandard products to reach consumers.

Europe requires ECE R90 homologation for replacement brake pads and shoes. Products must be tested and approved as performing within defined parameters relative to the OE part they replace. This creates a meaningful barrier to entry for low-quality manufacturers, raising the quality floor of the European aftermarket but also increasing costs and limiting competition.

This regulatory asymmetry explains why the North American aftermarket has more quality tiers and wider price variation than Europe, and why European aftermarket brands emphasize their R90 approval as a quality signal.

THE OE-AFTERMARKET CONNECTION

The two business models are more interconnected than they first appear:

Aftermarket pull is one of the primary reasons companies pursue low-margin OE business. Being specified as OE on a vehicle platform creates brand awareness and perceived quality validation that drives aftermarket sales years later when those vehicles need replacement parts. A brake pad manufacturer that's OE on a popular vehicle platform has a built-in marketing advantage in the aftermarket for that application.

OES (Original Equipment Service) programs allow OE suppliers to sell replacement parts through dealer networks at premium pricing. This is high-margin business—the same part sold at OE pricing to the assembly line might sell at three to four times that price through the dealer parts counter.

Technology transfer between OE and aftermarket is not strictly one-directional. Some innovations developed for new vehicles gradually become available in replacement parts, but the reverse also occurs. In the case of copper-free formulations, most aftermarket companies had zero-copper pads available well before OE adoption was widespread. Aftermarket formulations are markedly different from OE formulations and have completely different design factors, so the idea that aftermarket simply follows OE is an oversimplification.

THE ECONOMICS

Aftermarket margins are substantially higher than OE at every level. Where OE operating margins run in the low single digits, aftermarket manufacturers often achieve several times that. Distributors and installers add their own margins, creating the cumulative markup that determines final pricing.

This margin structure exists because the aftermarket is a convenience and availability business rather than a lowest-bidder business. An installer needs the right part today, not three months from now at the lowest possible price. The WD that can deliver overnight, the brand the installer trusts, and the catalog coverage that ensures the right application is in stock all command margin premiums that would be impossible in OE's competitive bidding environment.

For a deeper examination of aftermarket competitive strategy—margin economics at each distribution level, pricing tactics, brand building, coverage decisions, and the impact of e-commerce—see Chapter 28: Winning in the Aftermarket.

KEY TAKEAWAYS

- The aftermarket is fundamentally different from OE: purchasing decisions are made by installers and consumers in real time, not by automakers years in advance.
- Distribution flows through multiple levels—manufacturer, warehouse distributor, jobber, installer—each adding value and margin. Buying groups aggregate independent purchasing power and wield significant influence over brand selection.
- Catalog data infrastructure (FMSI D-numbers, ACES, PIES) and application coverage are critical competitive dimensions that determine whether products are visible and available to buyers.
- Quality spans from OE-equivalent premium to unregulated economy tiers, with North America's open market creating wider variation than Europe's ECE R90 homologation requirement.
- OE and aftermarket businesses are interconnected: OE specification drives aftermarket pull, dealer OES programs generate premium margins, and technology innovations transfer from new vehicles to replacement parts.

10. THE TIER STRUCTURE: HOW COMPONENTS REACH THE VEHICLE

Chapters 6 and 9 explain how OE and aftermarket business operate. This chapter explains the manufacturing tier structure that sits behind both—the layered system of suppliers, sub-suppliers, and raw material providers that turns iron ore and chemical compounds into the brake components that ship to assembly plants and distribution warehouses worldwide.

Understanding your company's position in this structure—and the positions of your customers, competitors, and suppliers—is essential for navigating business relationships and career decisions. Where you sit in the tiers shapes everything from your margin structure to your daily interactions.

THE TIERS

OEMs (Original Equipment Manufacturers) are the vehicle companies whose badges appear on cars and trucks. They design vehicles, assemble them, and sell them to consumers. OEMs rarely manufacture brake components themselves; instead, they source from their supply base. In braking, the OEM's engineering team defines system-level requirements—stopping distance, pedal feel, weight targets, packaging constraints—and the supply chain delivers against those specifications.

FIGURE 6: Brake Industry Supply Chain

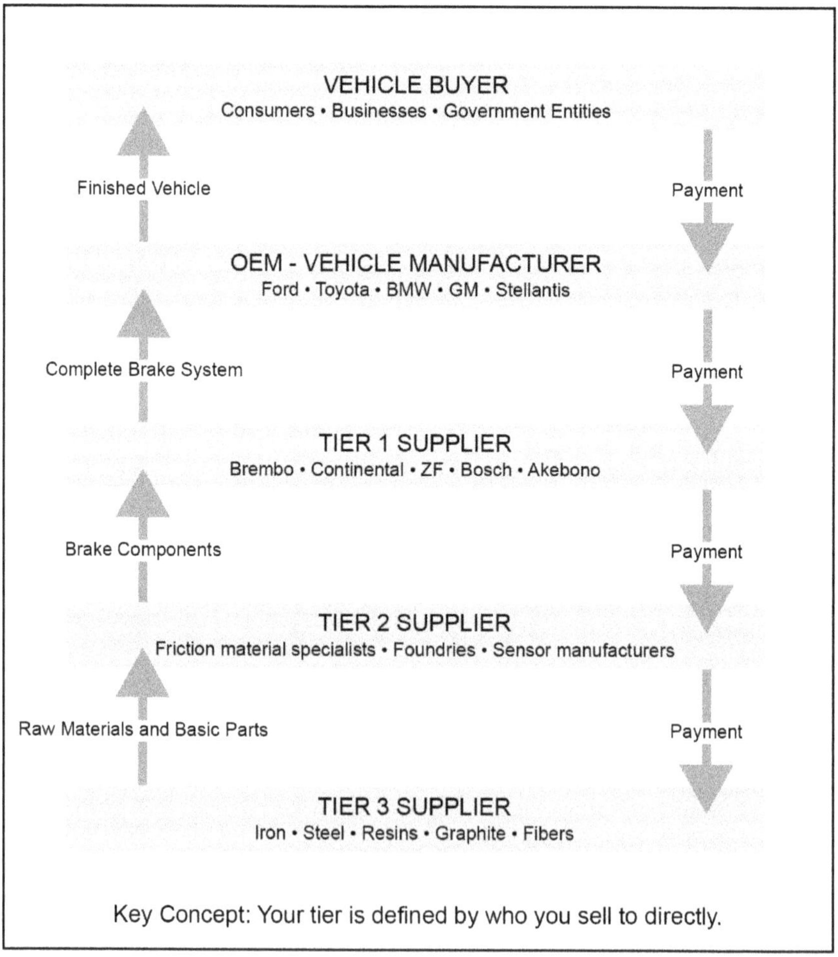

Tier 1 Suppliers sell directly to OEMs. In braking, Tier 1 suppliers typically provide complete brake systems or major assemblies—a full corner module including caliper, rotor, pads, bracket, and hardware, or an integrated electronic braking unit. They manage system-level engineering, coordinate the components sourced from their own supply base, and take responsibility for warranty and quality at the system level. Tier 1 suppliers often have resident engineers at OEM facilities and participate in vehicle development from early concept stages. Companies like Brembo, ZF, AUMOVIO, Bosch, Aisin (through its ADVICS subsidiary), Hyundai Mobis, and Mando (HL Mando) operate primarily at Tier 1.

Tier 2 Suppliers sell to Tier 1 suppliers rather than directly to OEMs. A Tier 2 might supply brake pads to a Tier 1 that assembles complete corner modules, provide castings that a Tier 1 machines into finished rotors, or manufacture the shims and hardware that a Tier 1 integrates into its system. Tier 2 suppliers typically have narrower product scope but deeper specialization in their area of expertise.

Tier 3 and Beyond includes raw material suppliers, commodity component makers, and other suppliers further removed from the vehicle. This might include steel mills providing gray iron for castings, resin suppliers providing binders for friction materials, abrasive manufacturers, rubber compound suppliers for seals and boots, or fastener companies. These suppliers often serve multiple industries and may not think of themselves as "brake industry" companies at all.

THE BLURRED LINES

In practice, the tier structure isn't as clean as the textbook version suggests.

Companies span multiple tiers. A supplier is Tier 1 when it sells complete brake systems to Tesla, for example, but it's effectively Tier 2 when another Tier 1 integrator sources calipers as components from them. A friction material company might be Tier 1 when it sells replacement pads under its own brand through the aftermarket, but Tier 2 when it supplies pads to a system integrator for OE programs.

Direct Tier 2-to-OEM relationships exist. Some OEMs maintain direct relationships with critical Tier 2 suppliers even though the commercial contract flows through a Tier 1. The OEM might approve (or mandate) specific friction material formulations or casting sources, effectively controlling Tier 2 selection while leaving the Tier 1 responsible for logistics and quality management.

Vertical integration complicates the picture. Large brake companies often own capabilities across multiple tiers. A company that casts its own rotors, formulates its own friction materials, manufactures its own calipers, and assembles complete systems has essentially internalized Tiers 1 through 3. This integration can be a competitive advantage—tighter control over quality and supply chain—or a liability if internal operations aren't cost-competitive with the open market.

WHAT TIER POSITION MEANS FOR THE BUSINESS

Your company's tier position shapes the business in ways that aren't always obvious:

Customer concentration increases as you move up the tiers. A Tier 1 supplier might have five to ten OEM customers. A Tier 3 raw material supplier might sell to hundreds of companies across multiple industries. This means Tier 1 companies live and die by a handful of relationships, while lower-tier suppliers have more diversified risk but less pricing power with any individual customer.

Engineering requirements are highest at Tier 1, where you're responsible for system-level performance, integration with the vehicle, and managing the complexity of multiple component interactions. Tier 2 engineering is deep but narrow—you're optimizing your component, not the system. Tier 3 is primarily process engineering and material science.

Quality accountability cascades downward but concentrates upward. When a brake system fails in the field, the OEM's first call goes to the Tier 1 supplier, regardless of whether the root cause is a Tier 2 component or a Tier 3 raw material. The Tier 1 carries the warranty exposure and the quality scorecard impact, then works backward through the supply chain to identify and resolve the source. This asymmetric accountability is one reason Tier 1 suppliers invest heavily in incoming inspection and supplier quality management.

Margin structures reflect the value-add at each level. Tier 1 commands higher absolute margins for system integration and customer management but bears the highest commercial risk. Tier 2 margins are often more stable—you're selling a defined component at a negotiated price. Tier 3 margins are typically the thinnest, reflecting commodity market dynamics, but volumes can be enormous.

Switching costs increase at higher tiers. An OEM can't easily replace a Tier 1 brake system supplier mid-program—the validation investment is too large. But replacing a Tier 3 fastener supplier, while not trivial, involves far less requalification. This creates a strategic incentive for suppliers to move up the tier structure toward stickier customer relationships, even if margins per unit are lower.

NAVIGATING ACROSS TIERS

Career paths in the brake industry often cross tier boundaries. A quality engineer at a Tier 2 friction material company might move to a Tier 1 integrator to manage the supplier relationship from the other side. A sales executive at a Tier 1 might join an OEM purchasing team, bringing supplier-side perspective to the customer organization.

Understanding the tier structure helps you anticipate what the transition will feel like. Moving from Tier 2 to Tier 1 typically means more customer-facing responsibility, broader scope, and more political complexity. Moving from Tier 1 to an OEM means shifting from "we propose" to "we decide"—a fundamentally different power dynamic. Moving down the tiers usually means deeper technical specialization and less commercial complexity.

KEY TAKEAWAYS

- The brake industry operates through a tiered manufacturing supply chain: OEMs source from Tier 1 system integrators, who source components from Tier 2 specialists, who source materials from Tier 3 suppliers.
- In practice, tier boundaries are blurred—companies span multiple tiers, OEMs maintain direct relationships with critical sub-tier suppliers, and vertical integration means some companies internalize the entire chain.
- Tier position shapes customer concentration, engineering scope, quality accountability, margin structure, and switching costs in ways that affect both business strategy and individual career experience.
- Career paths frequently cross tier boundaries, and understanding what changes at each level helps professionals navigate moves effectively.

11.
INSIDE THE FACTORY

There's something visceral about walking through a brake manufacturing facility.

The foundry floors where cast iron rotors are poured. The pressing operations where friction material is bonded to backing plates under tremendous pressure. The machining centers where tolerances are measured in thousandths of an inch. The testing dynamometers that simulate millions of miles of braking cycles.

I've visited dozens of these facilities over the years, from small specialized operations to massive plants producing millions of components annually. What strikes me every time is the tension between craftsmanship and scale.

Brake components aren't particularly glamorous to manufacture. A rotor is essentially a precisely machined chunk of cast iron. A brake pad is friction material pressed onto a metal backing plate. Calipers are aluminum or iron housings with pistons and seals. The individual components are straightforward.

But the precision required is extraordinary. A rotor that's even the slightest thickness variation will cause brake pulsation. A pad with inconsistent friction material will perform unpredictably. A caliper seal that fails will leak fluid and compromise the entire system.

Quality control in brake manufacturing is demanding by necessity. Components are inspected at multiple stages. Finished products undergo testing that simulates real-world conditions. Statistical process controls monitor production runs for any drift from specification.

When I ask production folks what keeps them focused, the answer is always some variation of the same theme: "We know where these parts are going. We know what they have to do."

HOW ROTORS ARE MADE

Rotor manufacturing begins in the foundry. Gray iron is melted in cupola or induction furnaces at temperatures exceeding 2,700°F. The molten iron is carefully controlled for chemistry—carbon content typically runs 3.4–3.8%, silicon 1.8–2.4%, with trace amounts of manganese, copper, and other elements that fine-tune the metallurgical properties. Getting this chemistry right is critical: too much carbon produces brittle iron, too little produces iron that's too hard and prone to cracking.

The molten iron is poured into sand molds—either green sand (bonded with clay and water) or resin-bonded sand, depending on the application. Vented (ventilated) rotors require more complex core assemblies to create the internal cooling vanes that channel airflow through the disc. The casting cools, the sand is broken away, and the rough casting emerges.

From the foundry, castings move to the machine shop. CNC lathes turn the braking surfaces to precise flatness and parallelism. Dimensional tolerances are measured in thousandths of an inch—typically 0.0005" or less for thickness variation across the braking surface. The hat section (the center portion that bolts to the hub) is machined for bolt hole patterns and register fits. Some high-end rotors undergo additional operations: drilling, slotting, or surface treatments like zinc coating for corrosion protection.

Every rotor is checked for balance—an unbalanced rotor creates vibration. Metallurgical samples are pulled from production runs to verify microstructure and hardness. Some manufacturers use automated inspection systems with laser measurement to check every piece; others rely on statistical sampling with tight SPC (statistical process control) limits.

HOW BRAKE PADS ARE MADE

Pad manufacturing is a different process entirely—more chemistry lab than machine shop.

It starts with raw material preparation. The friction compound—typically 15 to 25 ingredients—is weighed and blended according to the formulation. Ingredients arrive as powders, fibers, and granules: phenolic resin binder, reinforcing fibers (aramid, steel, ceramic, glass), friction modifiers (graphite, metal sulfides, aluminum oxide, silicon carbide, to name a few), and fillers (barium sulfate, calcium carbonate, mica). The mixing process must be precise—ingredient ratios, mixing sequence, and blending time all affect the final product.

Meanwhile, steel backing plates are stamped from sheet steel, cleaned, and treated with an adhesive layer. Some are also fitted with shims, wear indicators, or anti-noise coatings before the friction material is applied.

The core manufacturing step is hot pressing. The charge weights of blended friction compound are placed into a mold cavity on top of the prepared backing plate, and a hydraulic press applies tremendous force—typically 1,000 to 3,000 psi—while heating the assembly to 300–350°F. Under this combination of heat and pressure, the phenolic resin melts, flows through the compound, and begins to crosslink, binding the ingredients together and bonding the friction material to the backing plate. Press cycles typically run 3 to 8 minutes depending on the compound.

After pressing, pads move to curing ovens where they bake for several hours at gradually increasing temperatures—often stepping from 300°F up to 500°F or higher. This post-cure completes the resin crosslinking reaction and drives off residual volatiles. Proper curing is essential: under-cured pads may outgas during initial use (causing brake fade), while over-cured pads can become brittle.

Finished pads are ground to final dimensions, and the friction surface may be scorched (a brief, high-temperature surface treatment that burns off the outermost resin layer, reducing initial break-in fade). Wear indicator slots are cut if specified. Each lot undergoes testing—typically hardness checks, dimensional verification, and friction testing on a small-scale dynamometer.

HOW CALIPERS ARE MADE

Caliper manufacturing depends on the type. Cast iron calipers—common on standard vehicles—follow a foundry-to-machining path similar to rotors. Aluminum calipers, prevalent on performance and luxury vehicles, are typically produced by gravity casting or low-pressure die casting, then extensively machined.

The machining operations on calipers are demanding. Piston bores must be honed to mirror-smooth finishes with tight diameter tolerances—the piston-to-bore clearance determines seal performance and fluid leakage risk. Mounting surfaces, pad guide channels, and bleeder ports are all precision-machined. Multi-piston calipers (four-piston, six-piston, or more on high-performance applications) multiply the number of critical bores.

After machining, calipers are cleaned, inspected, and assembled with pistons, seals, boots, bleeder screws, and mounting hardware. Each assembled caliper undergoes pressure testing—typically at 1.5 to 2 times the maximum operating pressure—to verify there are no leaks or structural weaknesses. Some manufacturers also perform a vacuum test to ensure seal integrity.

Forged aluminum calipers, used by manufacturers like Brembo for high-performance applications, start with aluminum billets forged under extreme pressure to create a dense, strong blank with superior metallurgical properties compared to casting. The forging is then machined to final form—a more expensive process but one that produces a lighter, stronger caliper.

THE GLOBAL SUPPLY CHAIN

The brake industry's supply chain spans the globe. Raw materials—iron, steel, aluminum, friction material ingredients—come from multiple continents. Component manufacturing is distributed across regions to serve local vehicle production. Testing and validation may happen in yet another location.

This global footprint creates both efficiencies and vulnerabilities. The COVID-19 pandemic exposed supply chain risks that the industry is still working to address. Semiconductor shortages—which affect electronic brake systems—disrupted vehicle production for years. Geopolitical tensions have companies reconsidering their sourcing strategies.

Through my publishing work, I connect with brake professionals in over 100 countries. The challenges they describe are remarkably consistent: pressure on costs, demand for performance, increasing complexity from electronic integration, and the need to build supply chain resilience while remaining competitive on price.

It's a competitive industry, but also a surprisingly collegial one. Engineers at competing companies often know each other, having met at conferences, worked together earlier in their careers, or collaborated on industry standards. Through my recruiting work, I've seen firsthand how talent moves through this industry—and how much institutional knowledge experienced professionals carry with them.

KEY TAKEAWAYS

- Rotor manufacturing is a foundry-to-machine-shop process: gray iron is melted, cast in sand molds, then CNC-machined to tolerances of 0.0005" or less for thickness variation.
- Brake pad manufacturing combines chemistry and mechanical engineering: 15–25 ingredients are precision-blended, hot-pressed at 1,000–3,000 psi, then oven-cured for hours to complete the resin crosslinking reaction.

- Caliper production demands mirror-smooth piston bore finishes and pressure testing at 1.5–2x maximum operating pressure to ensure zero leaks.
- Quality control is demanding by necessity because the consequences of brake failure are severe—components are inspected at multiple stages with statistical process controls.
- The global supply chain spans continents, creating both efficiencies and vulnerabilities exposed by pandemic-era disruptions.

12. TESTING AND VALIDATION

No brake component reaches a production vehicle without extensive testing. The validation process is rigorous, multi-layered, and designed to ensure that brakes perform reliably across every condition they might encounter in service.

THE TESTING HIERARCHY

Material testing happens first, at the ingredient level. Friction material developers test raw materials for consistency and evaluate new compounds for their thermal, mechanical, and friction properties. This is laboratory work—measuring coefficients of friction across temperature ranges, assessing wear rates, evaluating noise characteristics.

Component testing evaluates finished parts. Brake pads undergo dynamometer testing that simulates thousands of braking cycles under controlled conditions. Rotors are checked for dimensional accuracy, balance, and metallurgical properties. Calipers are pressure-tested and evaluated for durability.

System testing evaluates how components work together. A complete brake corner—caliper, pads, rotor, mounting hardware—is tested as an assembly. The focus shifts to system-level performance: stopping distance, pedal feel, noise and vibration, heat management.

Vehicle testing puts brakes through real-world conditions, though it is worth noting that because prototype vehicles are extremely expensive, more and more OEMs are moving brake validation into the laboratory, completing complex validation procedures on dynamometers rather than in vehicles. The FMVSS 135 standard still requires a vehicle for measuring stopping distance, but much of the development testing that once required fleet vehicles can now be done on the bench.

When vehicle testing does occur, test vehicles are driven through standardized procedures (regulatory compliance testing) and subjected to extreme conditions (hot-weather testing in the desert, cold-weather testing in arctic conditions, high-altitude testing in the mountains, wet-weather testing). Engineers evaluate not just stopping performance but noise, vibration, pedal feel, and durability.

KEY TEST PROCEDURES

FMVSS 135 (Federal Motor Vehicle Safety Standard 135) is the U.S. regulatory standard for passenger car braking systems. It specifies stopping distance requirements, pedal force limits, and performance under various conditions including partial system failure. Every vehicle sold in the U.S. must comply.

FMVSS 121 is the corresponding U.S. standard for air brake systems on heavy commercial vehicles, trailers, and buses. It specifies stopping distance requirements, brake system response times, and parking brake performance for vehicles equipped with air brakes. Chapter 14 covers commercial vehicle braking in detail.

ECE R13 and R13-H are the European equivalents, covering passenger vehicles and heavy vehicles respectively. These regulations specify performance requirements and testing procedures for vehicles sold in Europe and many other markets that follow European standards.

SAE J2521 is an industry-standard dynamometer test for brake noise evaluation. It subjects brake assemblies to a series of stops under varying conditions to evaluate squeal, groan, and other noise issues.

AK Master (SAE J2522, widely used throughout the world) is a comprehensive friction material evaluation procedure widely used in Europe, consisting of multiple test sequences that evaluate fade, recovery, speed sensitivity, and other characteristics.

Beyond full-assembly dynamometer testing, the industry relies on sample-level tests to characterize friction materials independent of other brake components.

The Chase Test (SAE J661) was originally developed by T.P. Chase while working for General Motors and Link Engineering Company in the mid-1940s. The test uses a one-square-inch sample of friction material rubbing against the interior of a precisely machined cast iron drum on a Chase Machine. Under closely controlled conditions, the sample is run through a duty cycle that exposes the material to a thermal history similar to in-service performance. The results yield a coefficient of friction value for normal and hot thermal conditions, expressed as a two-letter edge code. This code is required for AMECA certification and must be stamped on the edge of every brake friction material sold for OE and aftermarket use in the United States.

The Friction Assessment Sample Test (FAST) Machine was developed by Dr. Arne Anderson of the Ford Research Laboratories in the mid-1960s. It tests a half-square-inch sample of friction material through a 90-minute constant-force, constant-speed procedure, allowing temperature to rise from ambient to 660°F. The precise control of all test conditions ensures reproducibility, making the FAST Machine a common tool for friction material formulation development as well as quality control and product assurance testing.

Some of the companies providing world-class testing equipment and/or services include Horiba, Bruker, Link Engineering, Greening, Inc., and TecSA.

THE ROLE OF SIMULATION

Increasingly, physical testing is complemented by computer simulation. Finite element analysis models how components will behave under stress. Thermal simulations predict heat distribution and fade characteristics. Noise and vibration models help identify potential issues before parts are manufactured.

Simulation doesn't replace physical testing—regulators require actual test data for certification—but it accelerates development by identifying problems early and reducing the number of physical prototypes required. A modern brake development program might evaluate hundreds of virtual designs for every physical prototype built.

The testing and validation process adds significant time and cost to brake development, but it's non-negotiable. The consequences of brake failure are too severe to allow shortcuts.

KEY TAKEAWAYS

- Brake validation follows a hierarchy: material testing, component testing, system testing, and vehicle testing—each stage more comprehensive than the last.
- FMVSS 135 (U.S.) and ECE R13/R13-H (Europe) are the key regulatory standards specifying stopping distance and performance requirements.
- SAE J2521 is the industry-standard dynamometer test for brake noise evaluation; AK Master is widely used in Europe for friction material evaluation.
- Computer simulation accelerates development but doesn't replace physical testing—regulators require actual test data for certification.

PART III
APPLICATION SEGMENTS

13. MOTORSPORTS AND BRAKING

The story of Le Mans 1953 isn't just history—motorsports continues to be the proving ground where brake technology is pushed to its limits and innovations are battle-tested before reaching production vehicles.

THE ULTIMATE STRESS TEST

Racing places demands on brakes that far exceed anything encountered in normal driving. Consider Formula 1: cars decelerate from over 200 mph to under 50 mph in seconds, generating deceleration forces exceeding 5 Gs. Brake disc temperatures routinely surpass 1,800°F. Drivers may apply the brakes hundreds of times during a single race, with virtually no opportunity for the system to cool completely.

In endurance racing—Le Mans, Daytona, the Nürburgring 24 Hours—brakes must survive this punishment not for a single sprint but for hours on end, often through varying weather conditions and day-to-night temperature swings.

These extreme conditions compress decades of wear into a single event. Problems that might take years to manifest in normal driving reveal themselves in a matter of hours on the track.

TECHNOLOGY TRANSFER

Many technologies now standard on production vehicles were proven in motorsports first:

Carbon-ceramic brakes were developed for Formula 1 and prototype racing, where their weight savings and heat resistance provided competitive advantage. Today, they're offered on high-performance road cars from Ferrari, Porsche, Lamborghini, and others.

Brake-by-wire systems were refined in racing applications where precise electronic control enabled more sophisticated traction and stability management. The systems now enabling regenerative braking in EVs trace their lineage to the track.

Advanced friction materials often debut in motorsports. Formulations that prove themselves under racing stress become the basis for high-performance street applications.

Cooling solutions—brake ducts, vented rotors, heat-resistant coatings—were all pioneered in racing before becoming standard equipment.

BRAKE SUPPLIERS IN MOTORSPORTS

For brake companies, motorsports serves multiple purposes: a testbed for new technology, a showcase for engineering capability, and a marketing platform that builds brand cachet.

Brembo dominates Formula 1, supplying braking systems to the majority of teams. Their red calipers visible through the wheels of race cars create brand association that carries into the consumer market. When a performance car comes equipped with Brembo brakes, buyers understand what that signifies.

Other suppliers compete in different series—NASCAR, IndyCar, sports car racing, rally—each with distinct technical requirements. The engineering learnings from these programs inform production vehicle development for years afterward. Examples of these suppliers are PFC Brakes (USA), Alcon (UK), AP Racing (UK, now part of Brembo), and Pagid Racing (Germany).

The relationship between motorsports and production vehicles remains symbiotic: racing provides the extreme conditions to prove technology, while production vehicles provide the volume to make advanced technology economically viable.

KEY TAKEAWAYS

- Racing remains the ultimate brake stress test—F1 cars experience 5+ G deceleration with brake disc temperatures exceeding 1,800°F.
- Brembo dominates Formula 1, supplying the majority of teams; their red calipers have become synonymous with high performance.
- Innovations proven in motorsports—carbon-ceramic discs, multi-piston calipers, advanced friction compounds—eventually reach production vehicles.
- Motorsports experience is highly valued in brake industry careers, particularly for engineering and product development roles.

14. THE COMMERCIAL VEHICLE WORLD

When most people think about brakes, they think about passenger cars. But the commercial vehicle segment—trucks, buses, trailers, and specialized equipment—is a distinct world with different technology, different players, and different dynamics. The truck and trailer brakes that stop 80,000-pound vehicles operate on fundamentally different principles than passenger car brakes, creating a separate market with different suppliers, different challenges, and different career opportunities.

AIR BRAKE FUNDAMENTALS

The most fundamental difference: commercial vehicles predominantly use air brakes rather than hydraulic brakes. Understanding why explains much about commercial vehicle braking.

Why Air Instead of Hydraulic

Air braking became standard for heavy vehicles for several compelling reasons:

Fail-safe design: Air brakes are designed so that loss of pressure applies the brakes rather than releasing them. Spring brakes (parking brakes) are held released by air pressure; if pressure is lost, springs apply the brakes. This is critical for vehicles whose failure to stop could be catastrophic.

Supply availability: Compressed air is already present on commercial vehicles for other systems—suspension, accessories, and trailer supply. Leveraging the existing air system simplifies overall vehicle design.

FIGURE 7: Commercial Vehicle Air Brake System

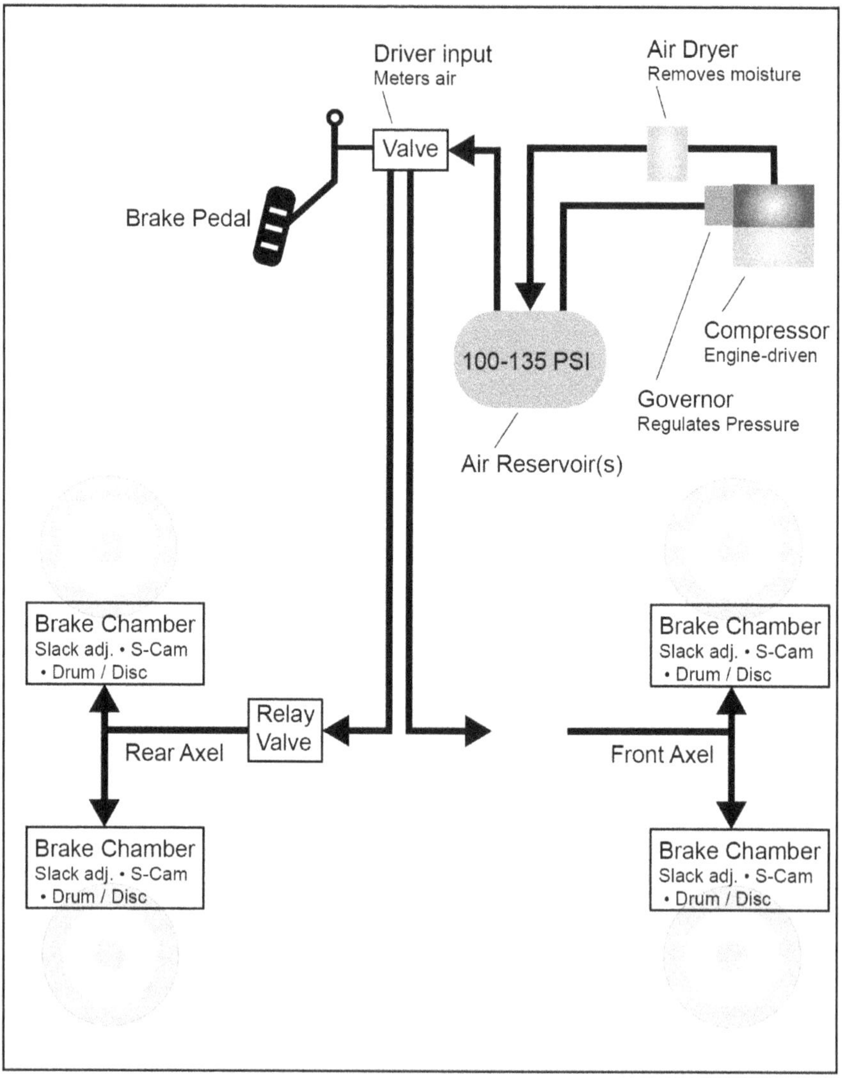

Scalability: Air systems can easily provide the force needed for heavy vehicle braking without the massive hydraulic cylinders that would otherwise be required.

Trailer compatibility: Air systems allow the tractor to supply air to trailer brakes through standardized gladhand connections. Hydraulic systems would require different trailer interface approaches.

Air Brake System Components

An air brake system includes many more components than a hydraulic system:

The compressor, driven by the vehicle engine, generates compressed air. Compressors must produce sufficient volume to operate brakes plus all other air-consuming systems.

Air dryers remove moisture from compressed air. Water in air systems causes corrosion, ice blockages in cold weather, and lubricant washout. Maintaining dry air is critical.

Reservoirs (air tanks) store compressed air. Multiple tanks with check valves ensure that a leak in one circuit doesn't drain the entire system.

Brake valves (foot valve, hand valve, trailer brake valve) control air flow to the brakes based on driver input.

Relay valves speed brake response by using control air to trigger local air supply rather than flowing all air through long lines from the driver's controls.

Quick release valves speed brake release by exhausting air locally rather than flowing back through the system.

Slack adjusters connect the air chamber pushrod to the S-cam shaft, converting linear pushrod motion to rotary S-cam motion and allowing adjustment for wear.

Brake chambers house diaphragms that convert air pressure to mechanical force.

FOUNDATION BRAKES: WHERE THE STOPPING HAPPENS

The foundation brake is the mechanical assembly that actually generates braking force. Commercial vehicles use several foundation brake types:

S-cam drum brakes are the most common in North America. An S-shaped cam rotates when the brake is applied, pushing brake shoes outward against the drum. Simple, robust, and well-understood, S-cam brakes handle most heavy-duty applications.

Wedge drum brakes use a wedge mechanism instead of an S-cam to force shoes against the drum. They're more compact and can be lower maintenance, but are less common than S-cam.

Air disc brakes (ADB) are increasingly common in commercial vehicle applications. Like passenger car disc brakes, they use calipers to clamp pads against a rotor—but actuated by air chambers rather than hydraulics.

The choice between drum and disc involves trade-offs:

Drum brakes are lower cost, well-proven, and technicians are universally familiar with service procedures. Heat dissipation can be challenging, and wet performance may suffer.

Disc brakes offer better heat dissipation, more consistent performance, easier inspection, and potentially longer pad life. They cost more initially and have historically faced resistance from fleets concerned about service infrastructure.

In practice, many heavy-duty tractor-trailer combinations today use disc brakes on the tractor and drum brakes on the trailer—a mixed configuration that reflects both the tractor's greater braking demands and the trailer market's slower adoption of disc technology.

Air disc brake adoption is accelerating, driven by multiple factors including safety benefits (shorter stopping distances), fleet operator preferences (they reduce maintenance costs and downtime), and the needs of electric commercial vehicles (which generate significant heat during regenerative braking). By the mid-2020s, air disc brakes accounted for a substantial and rapidly growing share of the heavy-duty vehicle brake market. Bendix, Knorr-Bremse, and others are actively developing and promoting air disc brake systems for applications that traditionally used drums. Many predict disc brakes will eventually dominate the commercial vehicle market as they have in passenger vehicles.

THE COMPETITIVE LANDSCAPE

The commercial vehicle brake market is a large and distinct segment within the global brake industry, with different leaders than the passenger car segment. The top players collectively hold the majority of the market, reflecting the technical complexity of air brake systems and the long-standing relationships with major truck manufacturers.

Knorr-Bremse (Germany) is the global leader in commercial vehicle and rail braking systems. Their truck division provides complete braking systems to major truck OEMs worldwide. In North America, they also own Bendix.

Bendix Commercial Vehicle Systems (USA) dominates the North American commercial vehicle market. A subsidiary of Knorr-Bremse since 2002, Bendix supplies foundation brakes, air dryers, ABS/stability systems, and driver assistance technologies to major truck and trailer manufacturers. They've recently streamlined their portfolio to focus on core braking competencies, launching new products like the ADB22X-LT air disc brake optimized for lighter-duty commercial applications.

WABCO was a major independent player until acquired by ZF in 2020. ZF now offers commercial vehicle braking through the legacy WABCO product lines, integrated with ZF's broader chassis systems capabilities.

Haldex (Sweden), now part of SAF-Holland following its 2022 acquisition, specializes in brake adjusters, air suspension, and specific component categories. Their automatic slack adjusters are found throughout the global commercial vehicle market.

Cummins Drivetrain and Braking Systems (formerly Meritor) provides axles, brakes, and drivetrain components to commercial vehicle OEMs.

Their drum and disc brakes serve both OE and aftermarket channels, with Meritor retained as the aftermarket brand.

TRAILER BRAKING

Trailers present unique braking challenges. They have no engine to provide vacuum or drive a compressor—all braking energy must come from the tractor through the air supply lines.

Trailer Air Systems

Tractors connect to trailers through two air lines:

The service line (blue gladhand) carries control air pressure proportional to driver brake application. This pressure doesn't directly apply the brakes; it signals trailer valves how much braking is demanded.

The supply line (red gladhand) provides the reservoir air that actually powers the trailer brakes. This line also serves as an emergency brake—if supply pressure drops (indicating disconnection or major leak), trailer spring brakes automatically apply.

Relay valves on the trailer use the control signal to modulate local air supply to the brakes, providing faster response than if all air had to flow from the tractor.

Trailer ABS

Trailer ABS (Anti-lock Braking System) has been required in North America since the late 1990s. The system uses wheel speed sensors to detect impending wheel lockup and modulates brake pressure to maintain control.

Trailer ABS communicates with the tractor through a PLC (Power Line Carrier) signal transmitted over the existing ABS lamp power wire, or through newer SAE J2497 data communication protocols.

ELECTRONIC INTEGRATION AND SAFETY TECHNOLOGY

Commercial vehicles are undergoing the same electronic transformation as passenger cars—perhaps even more dramatically, given the potential for autonomous trucking and advanced fleet management.

Electronic braking systems (EBS) that modulate air brake application electronically are becoming standard. These systems integrate with stability control, collision avoidance, and eventually autonomous driving functions.

Stability Systems

Roll stability control (RSC) uses body roll sensors to detect impending rollover and automatically applies brakes to reduce speed. Particularly important for tanker trucks and other high center-of-gravity vehicles.

Full stability control adds yaw control—detecting and correcting trailer swing or jackknife conditions by selectively applying individual wheel brakes.

Collision Mitigation

Active safety systems using radar and cameras can detect obstacles and automatically apply brakes:

Collision warning alerts the driver to impending collision.

Collision mitigation actively applies brakes when collision is imminent, reducing impact severity.

Automatic emergency braking (AEB) applies full braking force without driver input when collision is otherwise unavoidable.

These systems, once premium options, are increasingly standard equipment and will likely be mandated by regulation.

Regulatory Requirements

Commercial vehicle braking faces specific regulatory requirements:

FMVSS 121 specifies stopping distance and other requirements for air brake systems in the United States. Recent amendments have tightened stopping distance requirements, encouraging air disc brake adoption.

FMCSA regulations govern brake inspection, maintenance, and out-of-service criteria for commercial vehicles operating in interstate commerce.

The CVSA (Commercial Vehicle Safety Alliance) coordinates enforcement across jurisdictions and conducts the annual International Roadcheck inspection blitz.

Understanding these regulations is essential for anyone working in commercial vehicle braking—non-compliance can take vehicles off the road and create significant liability.

KEY TAKEAWAYS

- Air brakes dominate commercial vehicles because they're fail-safe (loss of pressure applies brakes) and can provide the force needed for heavy loads.
- S-cam drum brakes remain common, but air disc brakes are gaining share—now accounting for a substantial and growing portion of the heavy-duty market—driven by stopping distance regulations and total cost of ownership benefits.
- Key suppliers include Knorr-Bremse, Bendix, ZF/WABCO, Haldex, and Cummins Drivetrain and Braking Systems (formerly Meritor)—many of the same parent companies that supply the passenger vehicle market, though their commercial vehicle divisions operate as distinct businesses with specialized products and customer relationships.
- Trailer braking presents unique challenges since all braking energy must come from the tractor through air supply lines.
- Commercial vehicle braking offers distinct career paths requiring specialized knowledge of air systems, heavy-duty applications, and fleet customer dynamics.

PART IV
FORCES SHAPING THE FUTURE

15. THE REGULATORY PRESSURE COOKER

If you want to understand where the brake industry is heading, follow the regulations. More than any other factor, regulatory requirements shape R&D priorities, investment decisions, and competitive positioning.

Right now, three regulatory fronts are reshaping the industry: particulate emissions, copper elimination, and safety mandates.

EURO 7: THE PARTICULATE REVOLUTION

The most significant regulatory change in decades is Euro 7, which for the first time regulates non-exhaust emissions—including brake dust.

Here's what you need to know: Euro 7 sets limits on PM10 (particulate matter under 10 micrometers) from brake systems. Starting in late 2026 for new vehicle types and 2027 for all new vehicles, brakes must meet specific limits:

For most passenger cars: 7 mg/km of brake particulate emissions

For pure electric vehicles: 3 mg/km (lower because EVs use friction brakes less due to regenerative braking)

For large vans: **11 mg/km**

These limits will tighten further by 2030, and regulations will expand to cover particle number (PN) in addition to particle mass.

The implications are profound. The majority of brake systems currently in use don't meet these standards. Manufacturers must develop new friction formulations, new rotor coatings, and in some cases entirely new brake designs. One German company claims their coated brake discs can reduce dust emissions by up to 97%—but these solutions add cost and complexity.

The technology response includes low-emission friction materials (Tenneco, for example, has developed formulations using natural materials that have already been selected by European automakers), coated rotors that reduce wear, and brake dust capture systems that prevent particles from escaping into the environment.

China is developing similar standards (China 7), expected to follow shortly after Euro 7 implementation. This regulatory convergence means companies can't develop region-specific solutions—they need globally compliant technology.

COPPER-FREE MANDATES

The copper phase-out—covered in detail in Chapter 8—is now largely complete, with most new brake pad formulations meeting the sub-0.5% copper requirement. But for companies that lagged in developing copper-free solutions, market access is now restricted. It remains a clear example of how environmental regulations create both challenges and opportunities.

SUSTAINABILITY BEYOND COPPER: THE BROADENING CONVERSATION

The copper phase-out was the brake industry's first major sustainability reckoning, but the conversation has expanded well beyond a single material restriction. OEM purchasing teams now evaluate suppliers on comprehensive sustainability metrics, and the brake industry is responding across multiple fronts.

Circular economy principles are gaining traction in brake manufacturing. Traditionally, brake components follow a linear path: raw materials to production to installation to disposal. That model is shifting. Rotor remanufacturing—machining used rotors back to specification rather than melting them down—substantially reduces energy consumption compared to new casting. Some aftermarket companies have built business models around this concept. Caliper remanufacturing has been established for decades, but new processes are extending its reach to more complex assemblies.

Lifecycle assessment (LCA) is becoming a competitive differentiator. OEMs increasingly require suppliers to quantify the total environmental impact of their products—from raw material extraction through manufacturing, use phase, and end of life. For brake suppliers, this means tracking energy consumption per rotor cast, emissions per pad pressed, water usage in cooling processes, and the recyclability of finished products. Companies that can demonstrate lower lifecycle impact are winning platforms, particularly with European OEMs where ESG metrics directly influence sourcing decisions.

Recyclability is a growing concern, particularly for advanced materials. Traditional gray iron rotors are highly recyclable—scrap iron is one of the most recycled materials on earth. But carbon-ceramic rotors, increasingly common on performance and electric vehicles, present a recycling challenge. Carbon-silicon carbide composites are extremely durable but difficult to break down and reprocess. The industry is exploring grinding waste carbon-ceramic material for use as aggregate or filler, but no scalable recycling solution exists yet. As carbon-ceramic adoption grows beyond supercars into mainstream EVs, this gap will need to be addressed.

Brake dust capture systems represent perhaps the most visible sustainability innovation. Companies like Tallano Technologies, Mann+Hummel, and several Tier 1 suppliers have developed devices that capture brake particulate at the source—collecting dust before it enters the air. These range from passive collection systems using airflow dynamics to active systems with small fans and filters. Euro 7 regulations are accelerating adoption, and some OEMs have begun specifying brake dust capture as standard equipment on premium models. For brake suppliers, this creates both a new product category and a new engineering discipline.

The sustainability conversation is also reaching into manufacturing processes. Foundries are investing in electric arc furnaces powered by renewable energy. Friction material producers are reformulating to reduce reliance on mined minerals, substituting bio-based fibers and recycled content where performance allows. Packaging is shifting from single-use plastics to recyclable or returnable containers, particularly in the aftermarket channel where packaging waste is substantial.

For professionals entering or advancing in the brake industry, sustainability expertise is increasingly valuable. Roles in environmental compliance, LCA analysis, and sustainable product development are growing. Engineers who can optimize for both performance and environmental impact—who understand that reducing brake dust isn't just a regulatory exercise but a formulation challenge—will find themselves in high demand as these pressures intensify.

SAFETY MANDATES

Beyond emissions, safety regulations continue to expand. Automatic Emergency Braking (AEB) is becoming mandatory in more markets. The U.S. is implementing requirements for AEB on passenger vehicles, joining the EU and other regions that already mandate this technology.

For brake suppliers, these mandates drive demand for electronic braking systems capable of autonomous activation. The brake system becomes part of an integrated safety network, working with cameras, radar, and other sensors to prevent collisions. This is fundamentally different from traditional braking, where the driver initiates all braking action.

ABS and electronic stability control (ESC) are now standard on all new vehicles—ESC has been federally mandated in the U.S. since the 2012 model year. The regulatory frontier has moved to more advanced systems: collision avoidance, pedestrian detection, and eventually autonomous vehicle requirements that treat braking as one element of vehicle control.

KEY TAKEAWAYS

- Euro 7 represents the most significant regulatory change in decades—for the first time regulating non-exhaust particulate emissions from brake dust.
- The copper phase-out (less than 0.5% by 2025 in leading markets) forced complete reformulation of many friction materials.
- ADAS mandates (automatic emergency braking requirements) are driving brake system sophistication and electronic content.
- Regulatory compliance drives significant R&D investment—companies that anticipate requirements gain competitive advantage.

16.
THE ELECTRIC REVOLUTION

No conversation about the brake industry today is complete without addressing the elephant in the room: electric vehicles.

EVs are fundamentally changing how we think about braking, and the implications ripple through every corner of the industry.

The most significant change is regenerative braking. When you lift your foot off the accelerator in an EV, the electric motor reverses its function, becoming a generator that captures kinetic energy and feeds it back into the battery. This "regen" provides substantial deceleration—enough that many EV drivers rarely touch the brake pedal in normal driving conditions.

THE PARADOX OF LESS USE, SAME CRITICALITY

For brake component manufacturers, regenerative braking creates an interesting paradox: friction brakes are used less frequently, but they still need to perform flawlessly when called upon.

Consider the engineering challenges: A brake system that sits dormant for weeks, then suddenly needs to execute a panic stop. Rotors that may corrode from disuse in wet climates. Pads that must maintain friction characteristics despite infrequent engagement. These are different problems than traditional braking engineers faced.

I've talked to engineers developing solutions: "low-drag" calipers that minimize parasitic losses when brakes aren't being applied, corrosion-resistant rotor coatings (including the application of protective layers that prevent rust from forming during extended non-use), and friction formulations optimized for the unique duty cycles of EV applications.

There's also the weight factor. EVs are heavy—battery packs add significant mass—and that mass must be stopped when friction braking is needed. An electric truck might weigh substantially more than its internal combustion equivalent, placing greater demands on brake system capacity. At the same time, the EV market demands lightweight components to maximize range. It's a tension that brake engineers must navigate.

REGENERATIVE-FRICTION BLENDING

Modern EVs don't simply switch between regenerative and friction braking—they blend the two systems seamlessly. When you press the brake pedal, an electronic control unit determines how much deceleration to achieve through regeneration (capturing energy) and how much through friction (dissipating energy as heat).

This blending requires sophisticated software and precise actuators. It's why electro-hydraulic brake systems have proliferated in EVs—they provide the electronic controllability needed for smooth blending while retaining hydraulic backup for safety. EHB adoption has accelerated rapidly and is becoming standard equipment on EVs and hybrids (see Chapter 17 for detailed market data).

The transition to EVs isn't making brakes obsolete. It's making them different—and potentially more valuable as integrated electronic systems rather than standalone mechanical components. Chapter 17 explores how brake-by-wire technology is enabling this transformation, and Chapter 29 provides a complete technical reference for electronic brake systems.

KEY TAKEAWAYS

- Regenerative braking in EVs captures kinetic energy through the motor, providing the large majority of stopping power in normal driving and dramatically reducing friction brake wear.
- Reduced brake usage creates new challenges: corrosion from infrequent use, changed pad compound requirements, and longer replacement intervals affecting aftermarket volume.
- EVs are heavier than comparable ICE vehicles, requiring upgraded brake systems despite using friction brakes less frequently.
- The shift to EVs creates both threats (reduced aftermarket replacement volume) and opportunities (new technology requirements) for brake suppliers.

17.
THE TECHNOLOGY FRONTIER

The brake industry is in the midst of a technological transformation as significant as the shift from drum to disc brakes—and arguably more fundamental. We're moving from mechanical systems to software-defined systems, with profound implications for how brakes are designed, manufactured, and integrated into vehicles.

BRAKE-BY-WIRE: THE EVOLUTIONARY PATH

"Brake-by-wire" refers to replacing mechanical or hydraulic connections between the brake pedal and the wheel brakes with electronic signals. But the term covers a spectrum of technologies with different levels of integration.

Electro-hydraulic brakes (EHB) represent the current mainstream approach. They retain hydraulic fluid for the final actuation at the wheel, but use electronic control to modulate pressure. The driver's pedal input is sensed electronically, and a motor-driven pump generates hydraulic pressure based on software commands. This approach offers many benefits of electronic control while maintaining the familiarity and proven safety of hydraulic actuation. Systems like Bosch's iBooster and AUMOVIO's MK C1 fall into this category.

EHB has achieved massive scale: tens of millions of vehicles globally are now equipped with brake-by-wire technology, with China leading adoption growth at rates that industry analysts describe as rapid. The technology enables regenerative braking integration, stability control, and automated emergency braking.

Electro-mechanical brakes (EMB) take the concept further, eliminating hydraulic fluid entirely. An electric motor at each wheel directly drives the brake caliper. This is "true" brake-by-wire—no fluid, no hydraulic lines, pure electronic control.

EMB promises significant advantages: lighter weight (no fluid, lines, or master cylinder), faster response (direct actuation rather than fluid pressure), precise individual wheel control, and easier integration with vehicle software systems. For autonomous vehicles, EMB offers the responsiveness and controllability that advanced driving systems require. (For a detailed technical explanation of EHB and EMB architectures, including sensor technologies and functional safety requirements, see Chapter 29.)

But EMB faces real challenges around fail-safe operation, braking force density, and regulatory frameworks written for hydraulic systems. China took a significant step—becoming the first major market to do so—by approving new standards in 2025 that provide a legal basis for EMB installation starting January 2026, and the industry is now working to clear remaining technical hurdles for mass production.

SOFTWARE-DEFINED BRAKING

Perhaps more significant than the hardware changes is the shift toward software-defined braking—systems where braking behavior is determined by algorithms that can be tuned, updated, and optimized over time.

Brembo's Sensify system exemplifies this direction. It modulates braking force at each wheel independently, in real-time, based on vehicle dynamics and driving conditions. The brake pedal feel itself can be adjusted through software. Over-the-air updates can refine braking characteristics without physical service visits.

For vehicle manufacturers, software-defined braking enables differentiation—a sporty brand can tune aggressive brake response, while a luxury brand can prioritize smoothness. For brake suppliers, it shifts value from hardware toward software and systems integration capability.

The engineers I talk to increasingly describe their work in software terms: algorithms, sensor fusion, cybersecurity. The brake industry is becoming a software industry, and the competitive landscape is shifting accordingly.

ADAS INTEGRATION

Advanced Driver Assistance Systems (ADAS) are transforming brakes from driver-initiated systems to autonomous actors. Automatic Emergency Braking can apply full brake force without any driver input if sensors detect an imminent collision. Adaptive cruise control modulates brakes to maintain distance from vehicles ahead.

This integration requires braking systems capable of autonomous activation—fast, precise, and reliable enough to be trusted with life-safety decisions. The brake system must communicate seamlessly with cameras, radar, lidar, and central vehicle computers.

The regulatory push is accelerating adoption: AEB systems are now installed in a majority of new U.S. passenger vehicles, and mandates are expanding globally. In the United States, NHTSA's FMVSS No. 127 (finalized in 2024) requires AEB and pedestrian AEB on all new passenger cars and light trucks by September 1, 2029. For brake suppliers, ADAS integration is no longer optional—it's table stakes for competing in the OE market.

THE AUTONOMOUS FUTURE

If EVs represent evolution, autonomous vehicles represent revolution.

The brake systems in self-driving vehicles need to do everything conventional brakes do, plus interface seamlessly with sophisticated computer systems that make decisions in milliseconds based on sensor data. They need redundancy—backups for the backups—because there's no human driver to take over if something goes wrong.

Brake-by-wire systems become essential in this context. They offer the precision and response speed that autonomous systems require, along with the ability to integrate directly with vehicle control computers. A hydraulic system with mechanical linkages simply can't respond as quickly or precisely as software controlling electric actuators.

The concept of "fail-operational" (versus merely "fail-safe") becomes critical. In a conventional vehicle, a brake failure can be addressed by the driver taking alternative action—downshifting, using the parking brake, steering away from danger. In a fully autonomous vehicle, the brake system itself must continue operating even after component failures. This drives requirements for redundant motors, redundant electronics, redundant power supplies, and sophisticated fault detection and management.

Whether fully autonomous vehicles become mainstream in five years or fifty, the technological foundations being laid now are reshaping the industry. The engineering challenge is clear: brake systems must evolve from components a driver activates into intelligent subsystems that a vehicle's computer trusts with life-safety decisions—every time, without exception. Brake companies that can't develop electronic integration and software capabilities risk being commoditized—relegated to supplying basic hardware while others capture value from the intelligent systems layer.

KEY TAKEAWAYS

- Brake-by-wire systems replace mechanical/hydraulic connections with electronic signals—electro-hydraulic brakes (EHB) are mainstream; electro-mechanical brakes (EMB) are emerging.
- Brake-by-wire enables precise, independent control of each wheel—critical for advanced driver assistance systems and autonomous vehicles.

- EMB is moving toward mass production, with China's 2026 regulatory approval marking a key milestone. The implications for system architecture and supplier capabilities are significant.
- Software increasingly defines brake system value—the companies that master software integration will lead the next era of the industry.
- Autonomous vehicles require "fail-operational" brake systems with full redundancy—backups for the backups—because there's no human driver to take over.
- Response time requirements tighten significantly for autonomous applications—systems need brakes that can react in milliseconds based on sensor data.

PART V
YOUR PLACE IN THE INDUSTRY

18.
TYPES OF BRAKE COMPANIES

The brake industry encompasses a diverse ecosystem of manufacturers, each occupying a distinct niche. Understanding these different company types—their business models, capabilities, and cultures—helps professionals navigate career decisions and business relationships more effectively.

INTEGRATED SYSTEM SUPPLIERS

These are the giants of the brake industry—companies that design and manufacture complete braking systems including calipers, rotors, pads, hydraulic components, and electronic controls. They typically supply directly to OEMs and have the engineering depth to develop systems from concept through production.

Characteristics:

- Full-system engineering capability from concept to production
- Global manufacturing footprint to serve OEM assembly plants
- Heavy investment in R&D and testing facilities
- Strong balance sheets to fund multi-year development programs
- Both OE and aftermarket presence

Career Implications: These companies offer broad career paths across engineering, manufacturing, quality, and commercial functions. Expect structured development programs, international assignment opportunities, and exposure to cutting-edge technology. The trade-off is that you may be a small cog in a large machine, with less visibility and autonomy than at smaller companies.

FRICTION MATERIAL SPECIALISTS

These companies focus exclusively on brake pads, linings, and related friction products. Their core competency is friction material formulation—the complex chemistry and engineering that determines how a brake pad performs across thousands of stopping events and temperature cycles.

Characteristics:

- Deep expertise in friction material science and tribology
- Extensive dynamometer testing capabilities
- Multiple friction formulations for different applications
- Strong presence in aftermarket through brand portfolios
- Often supply both OE and independent aftermarket channels

Career Implications: Friction specialists offer deep technical specialization for materials scientists and friction engineers. Commercial roles involve managing complex brand portfolios across OE and aftermarket. These companies often have strong cultures built around their technical heritage.

FOUNDRIES AND CASTING SPECIALISTS

Brake rotors and drums require specialized iron casting capabilities. Foundries that serve the brake industry have developed expertise in the specific metallurgy, casting processes, and machining required for brake applications.

Characteristics:

- Capital-intensive operations with significant fixed costs
- Specialized metallurgical expertise (gray iron, carbon-ceramic)

- High-volume production with tight dimensional tolerances
- Often located near raw material sources or in lower-cost regions
- Many serve multiple industries beyond automotive

Career Implications: Foundry careers involve metallurgy, process engineering, and high-volume manufacturing operations. These are often family-owned businesses with strong local roots. Career paths may be narrower but offer deep operational expertise.

CALIPER AND COMPONENT MANUFACTURERS

Some companies specialize in brake calipers, wheel cylinders, master cylinders, and other hydraulic components without producing friction materials. These require precision machining, hydraulic engineering, and often aluminum casting capabilities.

Characteristics:

- Precision machining and assembly operations
- Hydraulic system engineering expertise
- Often aluminum foundry capabilities for caliper bodies
- May serve both OE and performance/motorsports markets
- Quality requirements driven by safety-critical function

Career Implications: These companies offer strong opportunities in mechanical engineering, precision manufacturing, and quality. Performance and motorsports-focused manufacturers attract passionate enthusiasts who value the product over pure commercial metrics.

AFTERMARKET DISTRIBUTORS AND REMANUFACTURERS

Not all brake companies manufacture—some focus on distribution, private-label programs, or remanufacturing. These companies aggregate products from multiple manufacturers and serve the independent aftermarket through their distribution networks and brand relationships.

Characteristics:

- Strong distribution networks and customer relationships
- Private-label and house brand programs
- Inventory management and logistics expertise
- May add value through kitting, packaging, or light assembly
- Remanufacturers rebuild calipers, boosters, and other components

Career Implications: These companies offer commercial and operations careers rather than deep engineering roles. Strong opportunities in sales, marketing, supply chain, and e-commerce. Faster-paced environments with closer connection to end customers.

ELECTRONIC BRAKE SYSTEM SUPPLIERS

As braking becomes increasingly electronic—ABS, ESC, brake-by-wire—a category of suppliers has emerged that focuses on the electronic control units, sensors, and software that govern modern brake systems.

Characteristics:

- Software and electronics engineering focus
- Integration with vehicle-level systems (ADAS, autonomous driving)
- Rapid technology evolution requiring continuous investment
- Often part of larger automotive electronics suppliers
- Growing importance as vehicles become more automated

Career Implications: These roles require software engineering, controls engineering, and systems integration skills. Career paths may lead into broader automotive electronics or autonomous vehicle development. High demand for talent creates strong compensation and mobility.

KEY TAKEAWAYS

- Integrated system suppliers (Brembo, ZF, AUMOVIO) design complete brake systems; component specialists focus on specific products like friction materials or castings.
- OE-focused companies have different cultures, processes, and career paths than aftermarket-focused companies—understanding the distinction is crucial.
- Private equity ownership has become common in the brake industry, bringing both operational discipline and shorter investment horizons.
- Company type significantly affects career experience—a friction material specialist offers different development opportunities than an integrated system supplier.

19. CAREERS IN THE BRAKE INDUSTRY

Through my recruiting work, I've helped place executives and technical professionals at brake companies around the world. That experience has given me perspective on the career landscape in this industry—what roles exist, where the talent shortages are, and how people build successful careers.

THE FUNCTIONAL AREAS

Engineering remains the backbone of the industry. Mechanical engineers design brake components and systems. Materials engineers develop friction formulations. Test engineers validate performance across thousands of scenarios. Increasingly, software engineers are essential—developing the algorithms that control electronic braking systems, the interfaces that connect brakes to vehicle networks, and the diagnostic systems that monitor brake health.

Operations and Manufacturing encompass the people who actually make brake components—plant managers, production engineers, quality managers, supply chain professionals. As manufacturing becomes more automated and quality requirements more stringent, these roles demand both technical knowledge and operational excellence.

Sales and Business Development connect brake suppliers to their customers. In the OE world, this means long-cycle, relationship-intensive selling—understanding automaker programs years before launch, engineering solutions that meet specific requirements, negotiating multi-year contracts. In the aftermarket, it's about distribution partnerships, brand building, and serving a fragmented customer base.

Product Management bridges engineering and commercial functions—defining what products to develop, at what price points, for which customers. Product managers need both technical literacy and market insight.

WHERE THE OPPORTUNITIES ARE

Several trends are creating talent demand:

Electronics and Software expertise is in short supply. Traditional brake companies are competing with tech firms and other automotive sectors for engineers who understand embedded systems, sensor integration, and software development. Companies are building capabilities they didn't need a decade ago.

EV and Autonomous Vehicle Experience commands a premium. Engineers who have worked on regenerative braking integration, brake-by-wire systems, or ADAS-integrated braking bring valuable expertise that's difficult to develop internally.

Leadership Transitions create opportunities. The brake industry, like much of manufacturing, faces generational turnover. Experienced leaders are retiring, and companies need to develop or recruit their replacements. This creates paths for mid-career professionals to move into senior roles.

Geographic Expansion is opening new positions. As companies expand in Asia, establish technical centers in new regions, and diversify supply chains, they need people who can operate across cultures and build capabilities in new locations.

BUILDING A CAREER

The most successful brake industry professionals I've worked with share certain characteristics:

Technical depth combined with commercial awareness. Understanding both how brakes work and why customers buy them—the intersection of engineering and business—is valuable.

Cross-functional experience. People who have worked in multiple functions—engineering and sales, manufacturing and product management—develop broader perspective and more career options.

Global exposure. The brake industry is genuinely global. Experience working with or in different regions—understanding different customer requirements, regulatory environments, and business cultures—is increasingly important.

Continuous learning. The technology is changing rapidly. Professionals who stay current—whether through formal education, industry involvement, or self-directed learning—remain valuable even as the industry transforms.

For those entering the industry or considering it, the brake sector offers something that's increasingly rare: the opportunity to work on technology that genuinely matters. Every brake system shipped carries the responsibility of protecting lives. That purpose attracts a certain kind of professional—and creates a community that takes its work seriously.

WHAT THE WORK ACTUALLY LOOKS LIKE

Job titles in the brake industry vary wildly from company to company. One firm's "Brake Applications Engineer" is another's "Vehicle Integration Specialist" is another's "Chassis Systems Engineer." Rather than profile specific titles that may not exist at your company, this section describes the core work streams that exist at virtually every brake supplier. Focus on recognizing the work, not the job posting language.

Design and Simulation

This work stream covers the engineers who create brake components—designing rotors, calipers, brackets, anchor plates, and assemblies from initial concept through production release. The daily reality involves significant time in CAD (CATIA and NX are the dominant platforms in automotive) and finite element analysis software, balancing performance requirements against packaging constraints, weight targets, and manufacturability.

What surprises newcomers is how constrained the design space is. You're not designing from scratch—you're working within tight packaging envelopes dictated by wheel size, suspension geometry, and powertrain layout. A few millimeters of clearance can determine whether a design is feasible. Much of the intellectual challenge is creative problem-solving within rigid constraints rather than blue-sky design.

Design engineers interact heavily with their OEM customer's chassis engineering team, attending regular design reviews where every dimension is scrutinized. They also work closely with manufacturing—a beautiful design that can't be reliably produced at volume is worthless. The best design engineers develop strong intuition for what the factory can and can't do.

Skills that matter: 3D CAD proficiency, FEA for thermal and structural analysis, GD&T fluency, understanding of casting and machining processes, and the interpersonal skills to navigate design reviews with demanding customers.

Testing and Validation

Testing is where brake designs prove themselves—or don't. Test engineers plan and execute validation programs that can span months and involve dynamometer testing (simulating braking on a machine), vehicle-level testing (actual driving), environmental exposure, durability cycling, and noise evaluation.

A typical week might include writing test plans for a new program, analyzing data from overnight dynamometer runs, troubleshooting unexpected noise issues on a vehicle at the proving grounds, and reviewing test results with the customer's validation team. The work is highly data-intensive—modern brake dynamometers generate enormous datasets that require both statistical rigor and engineering judgment to interpret.

NVH (noise, vibration, and harshness) testing deserves special mention because it's where many brake programs get stuck. A brake system can meet every performance metric and still fail if it squeals, groans, or produces vibration that the customer considers unacceptable. NVH evaluation combines instrumented measurement with subjective assessment—someone has to drive the car and judge whether the noise is acceptable. Engineers who develop good "ears" for brake noise are valuable and relatively rare.

Skills that matter: test planning and DOE (design of experiments), data analysis and statistical methods, dynamometer operation and programming, vehicle evaluation skills, and the ability to translate test results into actionable engineering conclusions.

Quality and Process Engineering

This is the work stream most directly connected to Chapter 7's quality ecosystem. Quality engineers at brake suppliers live in the world of APQP, PPAP, FMEA, SPC, and 8D—the frameworks that govern how products are developed, approved, and monitored in production.

The daily work divides roughly into two modes. In development mode, quality engineers are building control plans, conducting measurement system analyses, running capability studies, and assembling PPAP packages for customer approval. In production mode, they're monitoring statistical process control charts, investigating non-conformances, leading root cause analysis when issues arise, and managing customer quality concerns.

When a quality escape happens—a defective part reaches the customer—the quality team's world compresses dramatically. An 8D investigation demands rapid containment (sorting suspect inventory, sometimes at the customer's facility), rigorous root cause analysis, and corrective actions that satisfy both the immediate problem and the systemic vulnerability that allowed it. These incidents are high-pressure, high-visibility, and consequential for the company's commercial standing.

Quality roles at brake suppliers carry more weight than at many other types of manufacturers because the products are safety-critical. The customer's SQE (Supplier Quality Engineer) has effective veto power over production decisions, and the quality manager's signature on a PPAP submission carries personal accountability.

Skills that matter: IATF 16949 and core quality tools (APQP, PPAP, FMEA, SPC, MSA), root cause analysis methodologies, statistical analysis, audit experience, and the ability to communicate technical issues clearly to both internal teams and customers.

Manufacturing and Operations

The people who run brake manufacturing plants manage a combination of heavy industrial processes (casting, forging, pressing, heat treatment) and precision operations (machining, assembly, automated inspection). A brake rotor starts as molten iron and ends as a component machined to tolerances measured in microns. Managing that transformation at volume—millions of parts per year—while maintaining quality, controlling costs, and keeping people safe is genuinely challenging.

Plant managers and operations leaders balance competing demands daily: production schedules that must hit delivery windows (OEM assembly lines don't wait), quality metrics that must stay within control limits, cost targets that tighten annually, and workforce management in an era where experienced manufacturing talent is hard to find.

Process engineers focus on optimizing specific manufacturing steps—improving cycle times, reducing scrap rates, implementing automation, and troubleshooting when processes drift. In friction material manufacturing, process engineering is particularly critical because small variations in mixing, pressing, or curing parameters can meaningfully affect pad performance.

Skills that matter: lean manufacturing and continuous improvement methodologies, understanding of specific manufacturing processes (casting, machining, pressing, coating), production planning and scheduling, workforce management, and safety management.

Sales and Program Management

Selling brake components to OEMs is nothing like consumer sales. Cycles are measured in years, not months. A program manager at a

Tier 1 brake supplier might spend three years nurturing a relationship before seeing a purchase order, then manage that program for another seven years through production life.

The daily reality combines technical discussion (understanding the OEM's requirements and explaining your capability), commercial negotiation (pricing, annual productivity improvements, tooling amortization), and relationship management (building trust with engineering, purchasing, and quality contacts at the customer). Program managers are the single point of accountability for everything the customer experiences—if testing is late, if quality slips, if costs overrun, the program manager owns the problem regardless of where it originated.

Aftermarket sales operates differently—shorter cycles, more accounts, and a focus on distribution relationships rather than direct OEM engagement. Aftermarket sales professionals need to understand installer and distributor economics, competitive positioning, and brand value in ways that OE salespeople don't.

Skills that matter: technical literacy sufficient to discuss brake engineering credibly, commercial acumen for pricing and contract negotiation, project management discipline, customer relationship skills, and the resilience to manage complex programs over multi-year timescales.

Applications Engineering

This function sits at the boundary between the supplier and the customer, and it's where some of the most interesting problem-solving happens. Applications engineers adapt the supplier's products to specific vehicle platforms—selecting the right friction material for a particular vehicle's weight, performance profile, and NVH requirements; optimizing caliper piston sizes for pedal feel; tuning brake balance between front and rear axles.

The work is inherently cross-functional. Applications engineers need enough design knowledge to understand what's feasible, enough testing knowledge to interpret validation data, enough manufacturing knowledge to know what's producible, and enough commercial awareness to understand what the customer is willing to pay for. They're often the first people to identify problems and the last people to sign off before production.

At many companies, applications engineering is where young engineers develop the broadest understanding of the brake business. It's excellent preparation for later moves into program management, product management, or senior engineering leadership.

Skills that matter: broad technical knowledge across brake system fundamentals, data analysis, customer communication, problem-solving under time pressure, and the ability to synthesize inputs from multiple disciplines into practical recommendations.

THE FORMULATOR: FRICTION MATERIAL'S MOST CRITICAL ROLE

If there's one role in the brake industry that exemplifies specialized expertise, it's the friction material formulator—also known as a compounder. These professionals develop the brake pad formulations that determine stopping power, noise characteristics, wear rates, and virtually every other performance attribute. In an industry full of specialized knowledge, formulators represent the deepest specialization of all.

A brake pad formulation might contain 15 to 25 or more distinct ingredients: binders that hold everything together, reinforcing fibers that provide structure, friction modifiers that tune the coefficient of friction, abrasives that maintain rotor surface condition, and fillers that affect everything from cost to thermal properties. The formulator's job is to balance all of these elements to meet specific performance targets.

But it's not as simple as following a recipe. Formulators must account for interactions between ingredients—adding more of one component changes how others behave. They must understand processing variables, since the same ingredients mixed differently, pressed at different pressures, or cured at different temperatures will yield different results. They must navigate application requirements that differ fundamentally between a family sedan, a police pursuit vehicle, a heavy truck, and a racing application. And they must work within regulatory constraints like copper content limits while managing cost pressures on raw materials.

Experienced formulators are among the hardest professionals to recruit in the entire automotive supply chain. It takes 10 to 15 years of hands-on experience to develop the intuition that distinguishes good formulators from exceptional ones. Much formulator expertise is tacit—learned by doing rather than from documentation. No university program specifically trains friction material formulators; most come from chemistry or materials science backgrounds and learn formulation on the job. The global pool of experienced formulators probably numbers in the hundreds, not thousands.

When I recruit for formulator positions, hiring managers consistently prioritize depth of experience over breadth, problem-solving track records that demonstrate real expertise, understanding of manufacturing as well as chemistry, NVH intuition (since noise is often the hardest attribute to formulate for), and documentation discipline that captures institutional knowledge.

Formulators enjoy several career advantages. Compensation runs meaningfully higher than comparably senior engineers in other brake industry roles—the supply-demand imbalance is that stark. Job security is strong because companies are reluctant to lose formulators and often go to considerable lengths to retain them. The work is intellectually engaging, with constant problem-solving that formulators consistently cite as why they stay in the field. And expertise transfers across borders, creating global mobility.

Formulator positions exist at friction material specialists like TMD Friction, Nisshinbo Brake (Japan), Akebono, ITT Motion Technologies, and Frasle Mobility; at integrated brake suppliers like Brembo, AUMOVIO, and ZF that maintain in-house friction development; at aftermarket manufacturers developing competitive replacement products; and at testing laboratories that support friction material development.

For those interested in formulation careers, the typical path involves a degree in chemistry, materials science, or chemical engineering; entry through adjacent roles like quality, testing, or process engineering at friction material companies; apprenticeship under experienced formulators; and patience—accepting that deep expertise takes time. The five-year mark is when most formulators start contributing independently; the ten-year mark is when they become truly valuable; the fifteen-year mark is when they become irreplaceable.

KEY TAKEAWAYS

- Engineering (mechanical, materials, test, software) remains the backbone of the industry, but commercial roles in sales, marketing, and product management are equally critical.
- Software engineering is the fastest-growing skill set as brake systems become electronically controlled and integrated with vehicle networks.
- The industry faces talent shortages in specialized areas—experienced brake engineers command premium compensation.
- Careers often span multiple companies as professionals follow opportunities across the relatively small industry network.

20. GEOGRAPHIC MANUFACTURING HUBS

Brake manufacturing is concentrated in specific regions worldwide, each with distinct advantages and characteristics. Understanding why brakes are made where they are helps explain industry dynamics and informs career decisions about location.

WHY LOCATION MATTERS

Several factors drive brake manufacturing location decisions:

Proximity to OEM Assembly: Brake systems are heavy and bulky relative to their value, making shipping expensive. Just-in-time delivery requirements mean suppliers locate near OEM assembly plants. When an automaker builds a plant, brake suppliers often follow.

Labor Costs and Skills: Brake manufacturing involves both labor-intensive processes (pad production, assembly) and highly automated operations (precision machining). Different regions offer different cost/skill tradeoffs.

Raw Material Access: Foundries benefit from proximity to scrap iron sources. Friction material plants need access to diverse raw materials from graphite to aramid fibers.

Trade Policy: Tariffs, local content requirements, and trade agreements influence where companies locate manufacturing. Regional production often avoids cross-border duties.

NORTH AMERICA

North American brake manufacturing clusters around the automotive manufacturing heartland, with concentrations in the U.S. Midwest and Mexico.

U.S. Midwest: Michigan, Ohio, Indiana, and Kentucky host numerous brake facilities serving Detroit OEMs and transplant assembly plants. Strong engineering talent pool from automotive heritage. Higher labor costs offset by proximity to customers and established supplier networks.

Mexico: Growing brake manufacturing presence, particularly in northern states near the U.S. border. Lower labor costs while maintaining USMCA trade benefits. Many companies operate paired facilities—engineering and prototyping in the U.S., production in Mexico.

Canada: Smaller brake manufacturing footprint but integrated into North American supply chains. Strong in commercial vehicle components.

EUROPE

Europe maintains significant brake manufacturing despite cost pressures, driven by OEM proximity and technical heritage.

Germany: The heart of European brake engineering. Home to major supplier headquarters and R&D centers. Premium vehicle OEMs demand local engineering support. Higher costs but unmatched technical depth.

Italy: Strong in performance and high-end brake systems, reflecting the Italian automotive heritage. Design and engineering excellence with growing cost pressures.

Eastern Europe: Czech Republic, Poland, and neighboring countries have attracted brake manufacturing seeking lower costs while remaining within EU trade zone. Growing technical capabilities beyond pure cost play.

Spain and UK: Established brake manufacturing bases serving regional OEMs. Brexit has complicated UK supply chain integration.

ASIA

Asia represents both the largest growth market and an increasingly important manufacturing base for global supply.

Japan: Sophisticated brake manufacturing with strong ties to Japanese OEMs. Known for quality systems and continuous improvement culture. Serves as R&D hub for Asian operations of global suppliers.

China: Massive and rapidly growing brake manufacturing base. Initially focused on cost-competitive products for domestic market, increasingly capable of serving global OEMs and export markets. Quality varies widely but top manufacturers match global standards.

South Korea: Strong brake industry serving Korean OEMs with growing export capability. Known for rapid response and flexibility.

INDIA: THE EMERGING POWERHOUSE

India deserves special attention. What was once a peripheral market for global brake suppliers has become one of the industry's most dynamic growth stories—both as a domestic market and as a manufacturing hub for global supply.

The domestic market is substantial and accelerating. India is now the world's third-largest automotive market by volume, producing over five million passenger vehicles and nearly one million commercial vehicles annually. The government's push toward vehicle safety—mandating ABS on two-wheelers, dual airbags, and electronic stability control—is driving demand for more sophisticated braking systems. The two-wheeler segment alone represents a massive opportunity: India produces over 20 million two-wheelers annually, and the shift from drum to disc brakes on these vehicles is creating new volume for brake suppliers at every tier.

Several Indian companies have grown from domestic suppliers into globally relevant players. Brakes India (part of the TSF Group, formerly a joint venture with ZF until 2021) is one of the largest brake manufacturers in Asia, producing foundation brakes, calipers, drums, and rotors for both domestic and export markets. ASK Automotive has built a strong position in two-wheeler and passenger vehicle brake assemblies, expanding from India into international markets. Masu Brakes supplies friction materials and brake assemblies to both Indian OEMs and global aftermarket channels. Rane Brake Lining, part of the Rane Group, has been producing friction materials for decades and holds significant domestic market share. Sundaram Brake Linings (part of the broader TVS Group heritage) and Automotive Axles Limited add further depth to India's brake manufacturing base.

Global suppliers have recognized India's importance. Brembo, AUMOVIO, Bosch, ZF, and Akebono all have Indian manufacturing operations, typically serving both the domestic market and using India as an export base for cost-competitive components. The combination of engineering talent, competitive labor costs, and a growing domestic market makes India attractive for both production and R&D investment. Several global brake companies have established engineering centers in Bangalore, Pune, and Chennai, leveraging India's deep pool of mechanical and software engineers.

India's role as a global export hub is expanding. Indian-manufactured rotors, drums, and friction materials are increasingly competitive in North American and European aftermarket channels. The cost advantage is significant—Indian foundries can produce rotors at substantially lower cost than U.S. or European alternatives—and quality capabilities have improved dramatically over the past decade. Companies like Sundaram Clayton and Menon Bearings have built export businesses shipping brake components to global Tier 1 suppliers and aftermarket distributors.

The challenges are real but manageable. Infrastructure limitations—inconsistent power supply, port congestion, and logistics complexity—add friction to operations. Quality consistency across the supply base varies more widely than in mature markets. Intellectual property protection concerns remain for some global companies. And the competitive landscape is intense: India has hundreds of small brake component manufacturers competing aggressively on price, which can pressure margins for companies trying to maintain quality standards.

For career purposes, India represents a growing opportunity. Engineering and manufacturing management roles are plentiful, and international experience in India is increasingly valued by global brake companies. Professionals who understand both the Indian market and global OEM requirements—and who can bridge the cultural and operational differences—are in particular demand as companies scale their Indian operations.

Southeast Asia: Thailand, in particular, serves as manufacturing hub for Japanese OEM transplants. Growing capabilities in Indonesia and Vietnam.

CAREER IMPLICATIONS OF GEOGRAPHY

Where you work in the brake industry significantly impacts your career:

- Engineering and R&D roles concentrate in Germany, Japan, and the U.S. Midwest
- Manufacturing management opportunities are broader geographically
- International experience is increasingly valued—consider rotational assignments
- Language skills (German, Japanese, Mandarin) open doors
- Cost-region experience (Mexico, China, Eastern Europe) builds operational credibility

KEY TAKEAWAYS

- Brake manufacturing locates near OEM assembly plants due to just-in-time delivery requirements and the cost of shipping heavy components.
- Major hubs include: Detroit area (North American OE), Germany (European engineering), Japan (Asian OE), and increasingly Mexico, China, and Eastern Europe.

- Location decisions also reflect labor costs, technical talent availability, and trade considerations—tariffs and local content requirements influence facility placement.
- Career geography matters—certain locations offer more opportunities in specific segments (Michigan for OE, California for aftermarket/performance).

21.
HOW TO EVALUATE A BRAKE COMPANY

Whether you're considering a job offer, evaluating a potential supplier, or analyzing a competitor, understanding how to assess a brake company's health and prospects is a valuable skill. This framework provides the questions to ask and signals to watch.

FINANCIAL HEALTH

Financial stability determines whether a company can invest in its future, weather downturns, and honor commitments to employees and customers.

Questions to Ask:

- Is the company profitable? Look beyond revenue to operating margins and free cash flow.
- What is the debt level? Highly leveraged companies face risk when markets soften.
- Who are the owners? Private equity ownership often brings cost pressure and shorter time horizons. Family ownership may offer stability but limit investment.
- How dependent is the company on a few customers? Concentration risk is real—losing a major OEM program can be devastating.

- What does the backlog look like? Awarded but not-yet-launched programs indicate future revenue.

Warning Signs:

- Delayed capital investments or hiring freezes
- Supplier payment terms stretching longer
- Frequent executive turnover, especially in finance
- Customer quality scorecards declining
- Rumors of sale or restructuring

CUSTOMER RELATIONSHIPS

In the brake industry, customer relationships are everything. Quality escapes, delivery failures, or engineering missteps can destroy years of relationship building.

Questions to Ask:

- Which OEMs does the company supply? Diversity across customers reduces risk.
- What is the quality scorecard history? Ask about PPM rates and warranty performance.
- How long have key customer relationships existed? Long tenure suggests trust.
- Has the company been awarded new business recently? Winning programs signals competitiveness.
- What do customers say? Industry reputation travels fast in a small community.

Warning Signs:

- Lost programs or customers not being replaced
- Quality containment actions or customer-mandated inspections
- Revolving door in customer-facing roles
- Pricing pressure leading to cuts in engineering or quality resources

TECHNOLOGY AND ENGINEERING

Brake technology is evolving rapidly. Companies that fail to invest in next-generation capabilities risk obsolescence.

Questions to Ask:

- What percentage of revenue is invested in R&D?
- Does the company have capabilities in emerging areas (brake-by-wire, EV-specific solutions)?
- What testing facilities exist in-house vs. outsourced?
- How deep is the engineering bench? Experienced engineers or heavy reliance on junior staff?
- What patents or proprietary technology does the company hold?

Warning Signs:

- R&D budget cuts or engineering layoffs
- No clear strategy for EV transition
- Aging facilities and equipment
- Key engineers departing
- Over-reliance on customer-directed engineering vs. innovation

MANUFACTURING OPERATIONS

Operational excellence separates successful brake manufacturers from struggling ones. Quality, delivery, and cost all depend on manufacturing capability.

Questions to Ask:

- What is the facility age and condition? Modern equipment enables quality and productivity.
- What quality certifications are held? IATF 16949 is table stakes; customer-specific certifications matter.
- How automated vs. labor-intensive are operations?
- What is the plant's delivery performance? On-time delivery should exceed 98%.
- How is continuous improvement embedded in the culture?

Warning Signs:

- High employee turnover, especially in skilled positions
- Safety incidents or OSHA issues
- Deferred maintenance visible in facility condition
- Inventory problems—either shortages or excessive work-in-process
- Expedited freight becoming routine rather than exceptional

CULTURE AND LEADERSHIP

Culture determines whether a company is a place you want to spend your career. It also predicts how the company will behave under pressure.

Questions to Ask:

- How long has the leadership team been in place? Stability suggests a functioning organization.
- What is employee tenure like? Long tenure suggests people want to stay.
- How does the company develop talent? Training programs, promotions from within?
- What happens when things go wrong? Blame culture vs. problem-solving culture?
- How are decisions made? Empowered teams vs. everything escalating to the top?

Warning Signs:

- Executive churn or recent mass departures
- Glassdoor reviews consistently negative on specific themes
- People reluctant to speak freely during interviews
- No clear answer to 'what makes this a good place to work?'
- Emphasis on surveillance and control rather than trust and accountability

KEY TAKEAWAYS

- Financial health indicators include profitability, cash flow, debt levels, and R&D investment—sustainable companies invest in their future even during downturns.
- Customer concentration is a key risk factor—companies dependent on one or two OEMs are vulnerable to program losses or customer financial problems.
- Technology positioning matters increasingly—evaluate whether a company is investing in electronic systems, software capability, and emerging technologies.
- Culture and leadership stability affect long-term career prospects—high turnover in leadership often signals deeper organizational issues.

↱ For a ready-to-use evaluation framework, see the **Company Evaluation Scorecard** in Appendix D: Career Tools.

22.
WHAT HIRING MANAGERS ACTUALLY LOOK FOR

After facilitating hundreds of brake industry placements, clear patterns emerge in what separates candidates who get hired from those who don't. Understanding these factors can help you position yourself more effectively—and avoid common mistakes that derail promising candidates.

TECHNICAL COMPETENCE IS TABLE STAKES

Having the technical skills for a role is necessary but not sufficient. Most candidates who reach the interview stage have adequate technical backgrounds. What differentiates them is everything else.

What Hiring Managers Assume: If you've made it to the interview, they believe you can probably do the technical aspects of the job. Your resume already established that. The interview is about everything else.

Where Candidates Stumble: Spending too much interview time on technical minutiae rather than demonstrating judgment, communication, and leadership capability. Yes, you need to prove technical credibility, but don't let it crowd out the factors that actually drive hiring decisions.

PROBLEM-SOLVING TRUMPS PROBLEM-DESCRIBING

The brake industry is full of problems—quality issues, customer complaints, engineering challenges, delivery failures. Hiring managers desperately want people who solve problems rather than just identify them.

What They're Looking For:

- Specific examples of problems you personally solved, not problems your team addressed
- Evidence that you drove to root cause rather than treating symptoms
- Willingness to make decisions with incomplete information
- Resourcefulness in getting things done despite obstacles

What Turns Them Off:

- Vague answers about 'we' did this without clarity on your role
- Blaming others—previous employers, coworkers, customers—for failures
- Inability to articulate what you specifically contributed
- Problems described without resolution: 'It was a mess when I left'

CUSTOMER ORIENTATION MATTERS ENORMOUSLY

Whether the role is customer-facing or not, hiring managers want people who understand that customers pay the bills. In the brake industry, customer relationships are hard-won and easily lost.

What They're Looking For:

- Experience managing customer relationships, especially difficult ones
- Understanding of what customers actually care about (often different from what you might assume)
- Willingness to deliver bad news professionally and take ownership
- Track record of recovering from quality or delivery problems

What Turns Them Off:

- Dismissive attitude toward customer concerns

- No stories involving customer interaction
- Inability to see situations from the customer's perspective
- Technical arrogance: 'The customer doesn't understand'

CULTURAL FIT IS REAL (BUT OFTEN MISUNDERSTOOD)

Cultural fit doesn't mean hiring people who look and think alike. It means hiring people who will thrive in the company's actual working environment and share its core values.

Questions Hiring Managers Are Really Asking:

- Will this person work well with the existing team?
- Can they operate effectively in our decision-making style (consensus vs. directive)?
- Do they have the resilience for our pace and pressure level?
- Will they represent us well with customers?
- Are they motivated by what we can actually offer?

How to Assess Fit (For Yourself): Pay attention during interviews. How do people treat each other? Is the pace frantic or measured? Do people seem engaged or going through the motions? Trust your instincts—if something feels off, it probably is.

THE QUESTIONS BEHIND THE QUESTIONS

Experienced interviewers ask seemingly simple questions that reveal important information:

'Walk me through your resume': They're assessing your narrative skills, self-awareness, and whether your career shows intentionality or randomness.

'Why are you leaving your current role?': They're listening for red flags—do you badmouth your employer? Are your reasons credible? Is there a pattern of short stints?

'Tell me about a failure': They want to see if you take ownership, what you learned, and whether you have the self-awareness to recognize your own mistakes.

'Where do you see yourself in five years?': They're assessing whether your ambitions align with what this role can offer and whether you'll stay long enough to contribute.

'What questions do you have for me?': Your questions reveal what you care about and how deeply you've thought about this opportunity.

THE DECIDING FACTOR: WOULD I WANT TO WORK WITH THIS PERSON?

At the end of every interview, hiring managers ask themselves a simple question: Would I want this person on my team? Would I want to be in a tough meeting with them? Would I trust them to represent us to a customer?

Technical skills can be developed. Industry knowledge can be learned. But fundamental qualities—integrity, work ethic, judgment, coachability—are much harder to change. Hiring managers are ultimately betting on the person, not just the skill set.

KEY TAKEAWAYS

- Technical competence is table stakes—what differentiates candidates is problem-solving approach, communication ability, and cultural fit.
- Industry experience matters more than in other sectors because the brake industry has unique requirements, relationships, and institutional knowledge.
- Hiring managers assess trajectory and potential, not just current capabilities—demonstrated learning ability and career progression patterns matter.

- Red flags include job-hopping without clear rationale, inability to discuss failures constructively, and speaking negatively about previous employers.

↳ Hiring managers: see the **Candidate Evaluation Matrix** in Appendix D: Career Tools for a structured scoring framework.

23.
INTERVIEW QUESTIONS YOU'LL FACE

Brake industry interviews follow predictable patterns. Understanding the questions you'll face—and what interviewers are really assessing—helps you prepare effectively. The following questions are organized by role type, with guidance on what strong answers demonstrate.

ENGINEERING ROLES

Technical Questions:

"Walk me through how you would approach a brake NVH issue that appeared in vehicle testing."

They're assessing your diagnostic methodology. Strong answers demonstrate systematic problem-solving: gathering data, isolating variables, understanding the physics, considering multiple root causes before jumping to solutions.

"Describe your experience with DFMEA/PFMEA processes."

They want to know if you understand risk-based thinking and can facilitate cross-functional analysis. Specific examples of FMEAs you've led and how they influenced design or process decisions make strong answers.

"How do you balance performance requirements against cost targets?"

Engineering isn't just about technical optimization—it's about making tradeoffs. Strong candidates demonstrate commercial awareness and the ability to find creative solutions that satisfy multiple constraints.

"Tell me about a product launch you supported. What went well? What would you do differently?"

Launches reveal how engineers perform under pressure and whether they learn from experience. Self-awareness about what could have been better is more impressive than claims of perfection.

QUALITY ROLES

Common Questions:

"Describe a significant quality escape and how you handled it."

Quality escapes happen. What matters is containment speed, root cause rigor, customer communication, and systemic fixes. Strong answers show ownership and learning, not finger-pointing.

"How do you handle disagreements with manufacturing about quality issues?"

Quality professionals must maintain standards while preserving relationships. They're looking for diplomacy, data-driven arguments, and escalation judgment.

"Walk me through your approach to a supplier quality audit."

Strong answers demonstrate systematic methodology, focus on both compliance and capability, and the interpersonal skills to identify issues without creating adversarial relationships.

"How would you rebuild a relationship with a customer after a quality problem?"

Customer recovery requires humility, transparency, and follow-through. Strong candidates have specific examples and understand that trust is rebuilt through actions, not words.

SALES AND COMMERCIAL ROLES

Common Questions:

"Tell me about a deal you lost. What happened?"

Everyone loses deals. What matters is honest assessment of why, what you learned, and how you've adjusted. Blaming factors outside your control is a red flag.

"How do you manage a customer relationship when your company has let them down?"

Sales people are often on the front line of problems they didn't create. Strong answers show composure, customer empathy, and ability to drive internal action without making excuses.

"Describe your approach to developing new business."

They want to understand your prospecting methodology, how you qualify opportunities, and whether you're strategic or just reactive.

"How do you work with internal teams (engineering, operations) to support customer needs?"

Sales effectiveness depends on internal influence. Strong candidates demonstrate collaborative approaches rather than throwing problems over the wall.

OPERATIONS AND MANUFACTURING ROLES

Common Questions:

"Tell me about a significant operational improvement you led."

They want specific metrics: productivity gains, cost reductions, quality improvements. Vague answers about 'continuous improvement culture' don't cut it.

"How do you handle a situation where you need to ship product but have quality concerns?"

This tests judgment and values. Strong answers demonstrate that safety and quality trump short-term delivery pressure, while also showing creativity in finding solutions.

"Describe your approach to developing your team."
Manufacturing leadership is fundamentally about people. Strong candidates have specific examples of developing talent, not just managing tasks.

"How do you manage relationships with union leadership?" (if applicable)

Labor relations require respect, consistency, and communication. Strong answers demonstrate fairness and problem-solving, not adversarial framing.

TESTING AND VALIDATION ROLES

Test engineers are hired for their methodology and judgment, not just their equipment skills. These questions probe whether candidates can interpret data, not just generate it.

"You're running an AK Master evaluation and see an unexpected friction drop on the third fade cycle. Walk me through your next steps."
This separates engineers who troubleshoot from those who just run procedures. Strong candidates verify the data first—checking thermocouple readings, confirming rotor condition, reviewing pressure traces—before concluding it's a material issue. They distinguish between a test artifact and a genuine performance problem, and they know that a single fade cycle result needs context from the full test sequence before anyone makes formulation decisions.

"How do you determine whether a noise event during dynamometer testing will translate to a real-world customer complaint?"
This is the question that separates experienced brake test engineers from general test technicians. Dynamometers can produce noise that

never occurs on the vehicle, and vehicles can produce noise that dynamometers miss entirely. Strong candidates understand the limitations of dyno-to-vehicle correlation—they talk about frequency ranges, the role of suspension and knuckle dynamics that the dyno doesn't replicate, temperature and humidity sensitivities, and why subjective vehicle evaluation still matters even with sophisticated instrumentation. The best answers demonstrate the judgment to recommend "this needs vehicle confirmation" rather than either over-reacting to every dyno squeal or dismissing noise data that doesn't fit expectations.

"Describe a situation where your test data contradicted what the design team expected. How did you handle it?"

Test engineers are the bearers of inconvenient truths. A rotor that was supposed to last 60,000 miles shows cracking at 30,000. A pad that passed every lab test squeals on the vehicle. The hiring manager wants to know whether you have the integrity to report bad news clearly and the communication skills to do it without creating an adversarial relationship with the engineers whose design didn't perform. The best answers show someone who presents data objectively, helps diagnose why the result diverged from predictions, and contributes to the solution rather than just delivering the problem.

"Walk me through how you would design a validation plan for a new friction material on a platform you've never worked with before."

This tests whether the candidate understands the full scope of brake validation or just the slice they've personally executed. Strong answers reference applicable standards (FMVSS 135, SAE J2522, customer-specific procedures), the sequence of screening tests before committing to full validation, how they'd identify the critical performance attributes for that specific application, and what vehicle-level testing is needed beyond dynamometer work. Candidates who immediately default to "I'd run the customer's standard test procedure" without discussing how they'd scope and prioritize are showing limited experience.

APPLICATIONS ENGINEERING ROLES

Applications engineers live at the intersection of supplier capability and customer need. Interviews probe whether candidates can think across disciplines and manage the tension between what the customer wants and what's technically and commercially feasible.

"A customer reports a noise complaint on a vehicle that passed all of your validation testing. How do you approach it?"
This is the quintessential applications engineering problem, and it reveals how a candidate thinks about systems. Weak answers focus immediately on the friction material. Strong answers start by gathering information—when does the noise occur, what are the conditions, is it reproducible, what's changed since validation—and then work through the system: hardware condition, rotor surface, shim attachment, caliper function, mounting hardware, even suspension and hub runout. They understand that a noise complaint on a validated system usually means something in the field differs from the validation conditions, and their job is to find what changed.

"You've been asked to recommend a friction material for a new platform. The vehicle is heavier than the outgoing model, the wheel package is the same size, and the customer wants improved NVH. How do you approach the selection?"
The constraints are deliberately in tension: heavier vehicle means more energy to manage in the same package, while the customer simultaneously wants quieter brakes. This tests whether the candidate understands tradeoffs at the system level. Strong answers discuss thermal capacity concerns with the fixed package, how they'd evaluate whether the existing rotor design can handle the increased energy, what friction material characteristics they'd prioritize (fade resistance and recovery without sacrificing noise performance), and how they'd set realistic expectations with the customer about what's achievable. Candidates who treat this as a simple "pick a pad from the catalog" exercise are revealing shallow experience.

"Tell me about a time you had to push back on a customer's specification because it wasn't technically feasible."
Applications engineers who only say "yes" eventually deliver "no" at the worst possible time—during validation or after launch. This question tests technical credibility and professional courage. Strong answers describe a specific situation, the technical basis for pushing back, how they communicated it (data-driven, with alternatives rather than just objections), and how the relationship was preserved or even strengthened by the honesty. The hiring manager is looking for someone the customer will trust precisely because they're willing to deliver unwelcome technical reality.

"How do you prioritize when you're supporting multiple vehicle platforms at different stages of development simultaneously?"
Applications engineers rarely have the luxury of focusing on one program at a time. They're typically juggling platforms at different lifecycle stages—one in early material selection, another approaching PPAP, a third dealing with a field issue in production. This question tests organizational skills and judgment about where to invest time. Strong answers demonstrate a framework for prioritization (customer impact, timeline urgency, commercial value) rather than just claiming to be good at multitasking.

PROGRAM MANAGEMENT ROLES

Program managers are the single throat to choke—they own the customer relationship and every deliverable attached to it. Interviews focus heavily on how candidates handle the pressure points that define the role: timeline crises, cross-functional coordination failures, and the commercial realities of multi-year programs.

"You're six months from Job 1 and validation testing has uncovered a noise issue that requires a friction material change. Walk me through how you manage this."

This is the nightmare scenario every brake program manager has faced or will face. Six months before launch, a material change triggers a cascade: new PPAP submission, revalidation testing, potential tooling changes, updated control plans, and a customer who needs to understand the timeline impact. Strong candidates demonstrate that they understand the full chain of consequences—not just the engineering fix but the quality system implications, the customer communication cadence, and the contingency planning if the replacement material needs iteration. They show calm under pressure and the ability to build a recovery plan that keeps the customer informed without overpromising.

"How do you manage a customer's expectations when your internal teams are telling you they can't meet the original timeline?"

Program managers sit between two worlds—the customer expects what was promised and the internal team is telling you it's not happening. This question reveals whether the candidate defaults to optimism (telling the customer it'll be fine while hoping internally), avoidance (letting the deadline pass without proactive communication), or professional management (early disclosure with a revised plan and options). Strong answers emphasize getting ahead of the problem: communicating early, presenting alternatives rather than just bad news, and maintaining credibility by never promising what you can't deliver.

"Describe how you've managed annual productivity commitments while your raw material costs were increasing."

Most OE brake programs include contractual annual cost reductions—typically a few percent per year. When raw material costs move the wrong direction, program managers are caught between a contractual obligation to reduce prices and an economic reality that makes reductions painful or impossible. This question tests commercial sophistication: Do they understand cost models well enough to identify real savings opportunities? Can they build a data-driven case for the customer when cost relief is needed? Have they negotiated material surcharge agreements? Strong answers show someone who manages the P&L proactively rather than absorbing margin erosion quietly until it becomes a crisis.

"Tell me about a program where the scope changed significantly after award. How did you handle it?"
Scope creep is the silent killer of brake program profitability. The customer adds a vehicle variant, changes a performance requirement, or requests additional testing—each change seems small in isolation but compounds into significant cost and timeline impact. This question tests documentation discipline and negotiation skill. Strong candidates describe how they tracked changes against the original scope, quantified the cost and timing impact, and negotiated appropriate adjustments with the customer rather than absorbing every change without pushback. Candidates who describe programs where "everything worked out fine" despite major scope changes are either exceptionally lucky or not recognizing the margin they left on the table.

FRICTION MATERIAL FORMULATOR ROLES

Formulator interviews are among the most technically demanding in the brake industry. Hiring managers are assessing depth of ingredient knowledge, diagnostic intuition, and whether a candidate truly understands the interactions between formulation, processing, and application performance—or merely follows established recipes.

"You've developed a formulation that meets all performance targets on the dynamometer but generates unacceptable noise on the customer's vehicle. What's your diagnostic approach?"
This reveals whether the formulator thinks in terms of chemistry alone or understands braking as a system. Weak candidates immediately reach for ingredient adjustments—adding more graphite, changing the rubber content. Strong candidates first investigate the system context: What are the noise conditions (temperature range, pressure, speed)? Is the shim appropriate for the frequency involved? What does the rotor

surface look like? How does the caliper compliance compare to the dyno fixture? They understand that a formulation interacts with every other component in the brake corner, and that the answer isn't always a recipe change. The best formulators know when the noise problem is theirs to solve and when it belongs to the hardware team.

"Walk me through how you would approach reformulating an existing product to eliminate copper while maintaining equivalent performance."

Copper-free reformulation is the defining technical challenge of the past decade in friction materials, and this question reveals real depth of understanding. Strong candidates don't just name substitute materials—they explain why copper worked in the first place (thermal transfer, friction stability at elevated temperatures, noise damping, rotor surface conditioning) and how different substitutes address each of those functions, often imperfectly. They discuss the iterative nature of the work—how replacing copper typically requires adjusting multiple other ingredients to rebalance the formulation. They acknowledge trade-offs honestly: which performance attributes were hardest to maintain, and where compromises had to be accepted. Candidates who describe the process as straightforward are either underestimating the challenge or overstating their experience.

"Describe a formulation where you had to make a significant tradeoff between competing performance attributes. What did you prioritize and why?"

Every formulation is a compromise. A pad that provides aggressive initial bite may generate more noise. A formulation optimized for low dust may sacrifice high-temperature performance. This question tests whether the candidate understands that the job is managing trade-offs, not achieving perfection across every attribute. Strong answers describe the specific application context that drove the prioritization—a luxury vehicle where NVH trumped fade resistance, a police pursuit package where stopping power at extreme temperatures was non-negotiable despite higher wear rates. The best formulators align their technical decisions with what the application actually demands rather than optimizing in a vacuum.

"How do you document your formulation development work so that someone else could understand your reasoning five years from now?"

This may seem like a mundane question, but it reveals one of the biggest concerns hiring managers have about formulators: institutional knowledge preservation. Much formulation expertise is tacit—learned through trial and error, stored in the formulator's head, and lost when they leave. Companies have been burned by formulators who kept everything in their notebooks and took the knowledge with them. Strong candidates describe systematic documentation practices—not just recording what they did but why they did it, what alternatives they considered, what the test results revealed, and what they would try differently next time. The best answers acknowledge that documentation discipline requires effort that doesn't produce immediate results but protects the company's investment in formulation development over the long term.

QUESTIONS TO ASK YOUR INTERVIEWER

The questions you ask reveal your priorities and preparation. Strong questions demonstrate genuine interest in the role and company, not just whether you'll get the job.

Insightful Questions:

- What does success look like in this role after the first year?
- What are the biggest challenges the team is facing right now?
- How would you describe the company's culture, and how has it evolved?
- What happened to the previous person in this role?
- What's the most important thing I could do in the first 90 days?

Questions to Avoid (Or Save for Later):

- Questions about salary, vacation, or benefits before an offer is made

- Questions that could be answered by basic research
- Questions that suggest you're already thinking about your next move
- Leading questions designed to show off rather than learn

KEY TAKEAWAYS

- Technical questions assess methodology more than specific answers—interviewers want to see systematic problem-solving approaches.
- Behavioral questions ("Tell me about a time when . . .") probe how you've handled challenges that predict future performance.
- Commercial role interviews focus on customer relationship experience, market knowledge, and ability to translate between technical and business perspectives.
- Always prepare questions to ask interviewers—thoughtful questions demonstrate engagement and help you evaluate the opportunity.

↦ For a complete preparation checklist, see the **Interview Preparation Checklist** in Appendix D: Career Tools.

24. NEGOTIATING COMPENSATION IN A NICHE INDUSTRY

Compensation negotiation in the brake industry follows patterns common to specialized manufacturing, but with nuances that matter. Understanding the dynamics helps you maximize your outcome while preserving relationships you'll need throughout your career.

UNDERSTANDING BRAKE INDUSTRY COMPENSATION PATTERNS

Before negotiating, understand how compensation varies across the brake industry. These patterns remain consistent even as specific numbers shift with inflation and market conditions.

Company Type Differentials

Compensation varies significantly by company type. Tier 1 OE suppliers—companies like Brembo, AUMOVIO, or Bosch—typically pay meaningfully more in base salary than aftermarket distributors for comparable roles. They compete for talent against other well-funded automotive suppliers and structure packages accordingly. A Director of Engineering at a global Tier 1 will out-earn their counterpart at a regional aftermarket company, though the aftermarket role may offer other advantages.

Private equity-backed companies often offer a different structure: lower base salary but higher bonus potential, equity participation, and significant upside if the investment plays out. This appeals to certain professionals but requires understanding realistic payout scenarios. Ask about historical bonus achievement, not just targets. Understand the equity structure, vesting schedule, and what happens in various exit scenarios.

Smaller independent companies and family-owned businesses may pay less than large multinationals but often offer faster advancement, broader scope, and more direct impact. A product manager at a mid-sized brake company might earn less than their counterpart at a Tier 1 but could own an entire product line rather than a narrow segment. Weigh total opportunity, not just immediate compensation.

Regional Patterns

Geography matters. Metro Detroit remains the center of North American automotive and commands the highest salaries for brake industry roles—but also the highest competition. Other automotive hubs like Nashville, the Carolinas, and parts of the Midwest offer strong opportunities at somewhat lower compensation, often offset by lower cost of living. Remote work has changed some dynamics, but most brake industry roles still require proximity to customers, engineering centers, or manufacturing operations.

International differences are substantial. European brake companies often offer lower base salaries than American counterparts but provide stronger benefits, more vacation, and better job security. Asian markets vary widely—Japanese companies tend toward conservative base pay with strong stability, while Chinese brake companies increasingly compete on total compensation to attract experienced talent from Western suppliers. If you're considering international roles, understand the full package, not just the headline number.

Role Relativities

Different functions command different premiums. Engineering roles—particularly those requiring specialized brake expertise like NVH, system integration, or friction materials—typically command the highest salaries at equivalent experience levels. A senior brake systems engineer with OE program experience is a scarce commodity. Sales and business development roles often have lower base salaries but significant bonus or commission potential; a strong year can exceed engineering compensation, while a weak year falls short.

Supply chain and operations leadership at corporate level falls in competitive range, with directors and VPs earning strong packages—especially those who can navigate the complexity of global brake supply chains. Product management and program management have grown in importance and compensation as brake systems become more integrated and software-defined. Quality leadership at corporate level—VP of Quality, Director of Global Quality—commands premiums when candidates bring OE customer relationships and proven track records.

How Brake Industry Compensation Stacks Up

Rather than publish specific dollar figures that become outdated quickly, here is how brake industry roles generally compare to broader benchmarks. Use the U.S. Bureau of Labor Statistics median for mechanical engineers (published annually at bls.gov) as your baseline, then apply these relativities.

Entry-level engineers (0–3 years) typically start at or slightly below the BLS median for mechanical engineers. Brake-specific knowledge—even at the entry level—commands the higher end; generic mechanical engineering backgrounds start lower.

Mid-level engineers and individual contributors (4–10 years) generally earn 1.2–1.5× the BLS median. Specialists in NVH, friction materials, or brake-by-wire systems trend toward the top of that range due to the limited talent pool.

Senior engineers and technical leads (10–20 years) reach 1.5–2× the median. Those with OE program launch experience or deep customer relationships command premiums beyond that.

Friction material formulators are the outlier. At mid-career they earn meaningfully more than general brake engineers at the same experience level, and senior formulators with 15+ years of proven formulation development push well beyond that. The scarcity premium is real—there simply are not many people who can do this work.

Sales and business development (mid-level) base compensation tracks roughly in line with mid-level engineering, but the structure is different: expect meaningful bonus and commission potential for strong performers, often a substantial portion of base.

Engineering management (Director level) typically earns 1.8–2.5× the BLS median. At the VP of Engineering level at a global Tier 1 supplier, compensation moves meaningfully higher still.

General management and plant leadership falls in a similar range to engineering directors. Plant managers at large OE facilities trend toward the upper end.

C-suite (VP and above at major suppliers) base compensation reaches 2.5–4× the engineering median, with total compensation packages substantially higher when including bonus, equity, and long-term incentives.

Across all levels, Tier 1 OE suppliers pay meaningfully more in base than aftermarket distributors for comparable roles. Private equity-backed companies often offer lower base but higher bonus potential and equity participation. Geography matters too—roles in high-cost metro areas pay more, but the gap has narrowed with remote and hybrid arrangements.

For current dollar figures, check the BLS Occupational Outlook Handbook, professional society salary surveys (SAE International publishes periodic data), and conversations with specialized recruiters who can provide real-time market intelligence for specific roles.

Experience Multipliers

Certain experience commands premiums beyond base qualifications. OE program experience—having launched brake systems with major automakers—adds significant value because that knowledge transfers directly. Customer relationships matter: a commercial director with established relationships at Ford or Volkswagen brings revenue potential that justifies premium compensation. Technical credibility with

OE engineering teams is equally valuable for those in engineering and product roles.

Technical specialties in growing areas multiply your value. Experience with brake-by-wire systems, EMB development, ADAS integration, or brake particulate emissions is increasingly valuable as the industry shifts. Professionals who bridge mechanical and software domains—understanding both traditional brake engineering and electronic controls—are particularly scarce and compensated accordingly.

UNDERSTANDING YOUR LEVERAGE

Negotiating power depends on factors you can assess before entering discussions:

Market Scarcity: How many people can do this job? Specialized roles—friction material chemists, brake NVH experts, experienced plant managers with brake background—command premiums. Generalist roles face more competition.

Timing: Urgency increases your leverage. If they've been searching for months, have a gap causing business pain, or have already lost their second-choice candidate, you're in a stronger position.

Alternatives: Your best leverage is a credible alternative. Another offer, a current role you're happy in, or genuine willingness to walk away changes the dynamic entirely.

Track Record: Candidates with demonstrated results—launched products, turned around quality, grew accounts—command premiums over those with experience but unclear impact.

COMPONENTS OF COMPENSATION

Total compensation in brake industry roles typically includes multiple components:

Base Salary: The foundation. Most negotiation focuses here, but it's not the only lever. Base salary typically has more flexibility for external hires than internal promotions due to equity concerns.

Annual Bonus: Many roles include bonus potential tied to company, business unit, and individual performance. Understand the target percentage, typical payout history, and what metrics drive it.

Sign-On Bonus: A one-time payment that can bridge gaps when base salary has limits. Particularly useful when you're leaving unvested compensation at your current employer.

Relocation: If you're moving, understand what's covered: moving expenses, temporary housing, home sale assistance, family travel. This can represent significant value.

Equity: Less common in private brake companies but relevant for public companies or private-equity-backed businesses. Understand vesting schedules and realistic valuation.

Benefits: Health insurance, retirement plans, vacation, and other benefits vary significantly. A rich benefits package can be worth thousands annually versus a bare-bones plan.

NEGOTIATION TACTICS THAT WORK

Know Your Number: Before negotiating, be clear on what you need versus what you want. What's your walk-away point? What would make you enthusiastic? Don't negotiate without this clarity.

Let Them Go First: When possible, let the employer state a range or offer before you name a number. You might be pleasantly surprised, or at least learn their constraints.

Negotiate the Whole Package: If base salary is constrained, explore other components. Sign-on bonus, title, start date flexibility, early review for increase, or additional vacation may have more flexibility.

Use Objective Criteria: Market data, cost of living differences, and your current compensation provide objective anchors. 'Based on market data for this role and location' is stronger than 'I want more.'

Be Professional, Not Adversarial: You'll be working with these people. Aggressive tactics that damage relationships aren't worth it. Firm but respectful negotiation is the goal.

COMMON MISTAKES TO AVOID

- Negotiating before you have an offer—this signals you're not serious about the role
- Accepting immediately without any negotiation—they expect some back-and-forth
- Lying about competing offers or current compensation—the industry is small and reputations matter
- Focusing only on base salary when other components have more flexibility
- Making ultimatums you're not prepared to follow through on
- Continuing to negotiate after accepting—this destroys trust before you start

RELOCATION CONSIDERATIONS

Many brake industry roles require relocation to manufacturing hubs. Factor these realities into your evaluation:

- Cost of living differences can dramatically affect purchasing power
- Spouse/partner employment opportunities vary significantly by location
- School quality, housing markets, and lifestyle factors affect family decisions
- Future career mobility—some locations offer more opportunities than others
- Relocation packages vary widely; understand what's covered before accepting

KEY TAKEAWAYS

- Negotiating leverage depends on market scarcity for your skills, company urgency, and your alternatives—assess these before entering discussions.

- The brake industry is small enough that reputation matters—aggressive negotiation tactics can backfire in a tight-knit community.
- Total compensation includes base salary, bonus, equity (if applicable), benefits, relocation, and non-monetary factors like role scope and growth potential.
- Get offers in writing and take time to evaluate—but don't overplay your hand with excessive counteroffers or delays.

For brake-specific compensation data by role, experience level, and function, Hagman Search publishes an annual Compensation & Talent Report with detailed salary ranges, market pressure ratings, and a talent scarcity index. It's available as a free download at www.hagmansearch.com.

25.
FIRST 90 DAYS: HOW TO ONBOARD YOURSELF

The first three months in a new brake industry role set the trajectory for your success. While companies vary in formal onboarding quality, the best performers take ownership of their own integration. Here's how to accelerate your effectiveness.

WEEK ONE: LISTEN MORE THAN YOU TALK

Your first week shapes impressions that last. Resist the urge to demonstrate your value immediately.

Focus On:

- Learning names, roles, and relationships—who influences whom?
- Understanding your boss's priorities and communication preferences
- Observing how decisions get made and how work flows
- Identifying the informal power structure alongside the org chart
- Setting up your workspace, systems access, and basic logistics

Avoid:

- Criticizing how things are done, even if asked for opinions
- Making commitments you don't yet know you can keep

- Comparing your new employer unfavorably to your previous one
- Trying to solve problems before you understand them

MONTH ONE: BUILD YOUR KNOWLEDGE FOUNDATION

Understand the Products:

- Get hands-on with the products you'll be working with
- Understand manufacturing processes, quality requirements, and cost structures
- Learn what differentiates your company's products from competitors
- Study warranty data and common failure modes

Understand the Customers:

- Learn who the key customers are and what they care about
- Review customer scorecards and understand current performance
- Identify upcoming program launches and potential risks
- If possible, attend customer meetings as an observer

Understand the Organization:

- Meet key people in adjacent functions—don't stay in your silo
- Learn the planning and budgeting cycles
- Understand how success is measured and rewarded
- Identify the biggest challenges the organization is facing

MONTH TWO: START CONTRIBUTING

By week five or six, you should be ready to add value while continuing to learn.

Look for Quick Wins:

- Problems you can solve with fresh perspective and your specific experience
- Process improvements that don't require permission or resources

- Knowledge gaps you can fill for the team
- Relationships you can build that bridge organizational divides

Establish Your Working Rhythm:

- Regular check-ins with your manager to ensure alignment
- Participation in key meetings appropriate to your role
- Building relationships with peers who will be long-term collaborators
- Developing your own information sources and feedback channels

MONTH THREE: DEMONSTRATE IMPACT

By the end of your first quarter, you should have tangible contributions to show.

Document Your Progress:

- Problems solved or improvements made
- Relationships built across the organization
- Knowledge gained that positions you for greater contribution
- Areas where you still need development or support

Seek Feedback:

- Schedule a 90-day review with your manager, even if not required
- Ask specific questions: What's going well? Where can I improve?
- Listen to feedback without defensiveness
- Adjust your approach based on what you learn

RELATIONSHIPS TO BUILD

Your network within the company determines your long-term effectiveness. Prioritize these relationships:

Your Manager: Obvious but essential. Understand their goals, pressures, and preferences. Make their job easier.

Your Manager's Peers: The people at your boss's level will shape opinions about you and control resources you'll need.

Key Customers (Internal or External): The people who depend on your work. Understanding their needs is essential for prioritization.

Operations/Manufacturing: If you're not in operations yourself, build relationships there. Understanding production realities makes you more effective.

Long-Tenured Employees: People who've been around know history, relationships, and unwritten rules. They can save you from costly mistakes.

Administrative Staff: Assistants, coordinators, and support staff make things happen. Treat them with respect and they'll help you navigate.

KEY TAKEAWAYS

- The first week should focus on listening and learning—resist the urge to demonstrate value immediately through bold pronouncements.
- By day 30, aim to understand your role's key metrics, your boss's priorities, the organizational dynamics, and where you can begin contributing.
- By day 90, you should have early wins, established key relationships, and a clear plan for the next six months.
- Document what you learn—institutional knowledge captured early becomes a valuable reference as you take on more responsibility.

26.
BUILDING A CAREER WHEN YOUR COMPANY GETS ACQUIRED

The brake industry has experienced significant consolidation over the past two decades, and this trend continues. If you build a career in this industry, there's a reasonable chance you'll experience at least one acquisition. Understanding how to navigate these transitions can mean the difference between career advancement and career disruption.

THE ACQUISITION TIMELINE

Acquisitions unfold in predictable phases, each with different implications for employees:

Announcement (Day 0): The deal is announced but not yet closed. You'll hear reassuring messages about 'business as usual' and 'valuing our people.' Take these with appropriate skepticism—decisions haven't been made yet because the buyer doesn't have access to detailed information.

Due Diligence and Closing (Months 1–6): Regulatory approvals are sought and detailed integration planning begins. The acquiring company is assessing talent, often through formal review processes. This is the critical period to make your value visible.

Integration (Months 6–24): Organizational structures are finalized, redundancies are identified, and synergies are captured. Most workforce reductions happen during this period. Uncertainty peaks.

Stabilization (Year 2+): The new normal emerges. Those who remain often have expanded responsibilities and opportunities as the combined organization takes shape.

REALITIES TO ACCEPT

Synergies Usually Mean Job Cuts: When acquirers promise 'synergies,' they typically mean combining functions and reducing headcount. Finance, HR, IT, and corporate functions are usually hit hardest. Operations and customer-facing roles are often more protected.

The Acquiring Company's Culture Usually Wins: Despite talk of 'best of both,' the buyer's processes, systems, and cultural norms typically prevail. Adapting to their way of working is usually more productive than resisting.

Decisions Are Often Made with Limited Information: Acquirers make retention decisions based on incomplete data—org charts, reputation, brief interactions. The best person doesn't always win. Politics and visibility matter.

Your Manager May Not Survive Either: Don't assume your manager will protect your position—they may be fighting for their own. Build relationships with decision-makers at the acquiring company directly.

STRATEGIES FOR SURVIVAL AND SUCCESS

Make Your Value Visible: Don't assume your contributions are known. Document your results, customer relationships, and institutional knowledge. When integration teams assess talent, you want your value to be obvious.

Control the Narrative: How others describe you matters enormously during acquisitions. Ensure your manager, peers, and customers would speak positively about your capabilities and attitude.

Be a Bridge, Not a Barrier: Acquirers value people who facilitate integration rather than resist it. Volunteer for integration teams. Help the new owners understand the business. Be seen as part of the solution.

Maintain Professionalism Regardless: Even if you're angry, scared, or frustrated, maintain professional behavior. The brake industry is small—you'll encounter these people again. Burning bridges is rarely worth it.

Build Relationships with New Decision-Makers: Identify who from the acquiring company will influence your future. Find reasons to interact, demonstrate competence, and build rapport. Don't wait to be evaluated—participate in the evaluation.

Keep Your Options Open: Update your resume, activate your network, and be open to recruiters. Having alternatives improves your negotiating position if you're asked to stay and provides security if you're not.

IF YOU'RE ASKED TO STAY

Being asked to remain post-acquisition is positive, but negotiate thoughtfully:

- Understand your new role, reporting structure, and expectations clearly
- Negotiate retention bonuses or guarantees if appropriate
- Assess whether the new situation offers genuine career opportunity
- Be realistic about cultural fit with the acquiring organization
- Get commitments in writing—verbal assurances mean little

IF YOU'RE LET GO

Separation isn't failure—it's a common outcome of industry consolidation:

- Negotiate your severance package—there's often more flexibility than initially offered
- Understand your benefits continuation and any non-compete restrictions
- Request references and letters of recommendation while relationships are warm

- Activate your network immediately—the industry knows your situation and many will help
- Consider whether this is an opportunity to make a change you've been contemplating

THE LONG VIEW

Acquisitions are disruptive but often create opportunity. Combined organizations need leaders. Expanded product portfolios require expertise. New customers demand support. Those who navigate transitions well often emerge in stronger positions than they held before.

The brake industry rewards resilience. Almost everyone who's built a long career has survived at least one acquisition, layoff, or major reorganization. These experiences, while painful, build skills and perspective that serve you throughout your career.

KEY TAKEAWAYS

- Acquisitions unfold in phases: announcement, regulatory approval, close, integration—each phase has different implications for employees.
- The 12–18 months following close typically determine your future: duplicate roles are eliminated, reporting structures change, and new leaders assess inherited talent.
- Strategies for surviving acquisitions include building relationships with acquirer leadership, volunteering for integration teams, and demonstrating flexibility.
- Sometimes the best move is recognizing when to leave—an acquisition that eliminates your growth path may be the prompt to pursue outside opportunities.

27. THE HIDDEN JOB MARKET

Industry estimates suggest that a large share of brake industry positions are never publicly posted. Understanding how to access this hidden job market is essential for career advancement.

WHY MOST POSITIONS AREN'T POSTED

Industry Size: The brake industry is relatively small and specialized. Hiring managers often know the talent pool personally or through one degree of separation.

Confidentiality: Many searches are confidential—replacing underperformers, restructuring, or strategic hires that competitors shouldn't know about.

Quality Concerns: Public postings generate high volumes of unqualified applicants. Hiring managers prefer referred candidates who come with endorsements.

Speed: Working through networks and recruiters is often faster than public posting processes.

HOW TO ACCESS HIDDEN OPPORTUNITIES

Build Your Network:

- Attend industry events and trade shows
- Join professional organizations
- Maintain relationships with former colleagues at other companies
- Connect with industry suppliers and customers—they often know about openings

Work with Specialized Recruiters: Executive recruiters who specialize in the automotive or brake industry have access to positions that are never publicly posted. They can also provide market intelligence about compensation trends and company cultures.

Be Visible:

- Publish technical articles or present at conferences
- Maintain an active LinkedIn presence with industry-relevant content
- Develop a reputation as a go-to expert in your specialty

WORKING WITH EXECUTIVE RECRUITERS

Be Responsive: When a recruiter reaches out, respond promptly even if you're not actively looking. Building relationships before you need them pays dividends.

Be Honest: Share your true career interests, compensation expectations, and any constraints. Recruiters work better when they understand your full picture.

Maintain Confidentiality: Don't discuss recruiter conversations with colleagues. Discretion builds trust and keeps you informed about future opportunities.

Provide Referrals: Even if an opportunity isn't right for you, suggesting qualified candidates builds reciprocity. Recruiters remember helpful contacts.

KEY TAKEAWAYS

- A large share of brake industry positions are never publicly posted—accessed through networks, recruiters, and direct relationships.
- The industry's small size means hiring managers often know candidates personally or through one degree of separation.
- Building visibility through industry events, professional associations, and publishing/speaking creates opportunities that job boards cannot.
- Specialized recruiters exist specifically because confidential searches and network-based hiring dominate this industry.

PART VI
INDUSTRY DEEP DIVES

28. WINNING IN THE AFTERMARKET

The aftermarket presents a different business model than OE—more fragmented, faster-moving, and ultimately more accessible to a broader range of competitors. Chapter 9 covered how the aftermarket operates: the distribution chain, quality tiers, catalog infrastructure, and regulatory landscape. This chapter goes deeper into competitive strategy—how companies actually win in this market.

THE AFTERMARKET VALUE CHAIN

Understanding how money flows through the aftermarket helps explain competitive dynamics.

The Margin Stack

Consider a brake pad that retails for $60. That price represents the accumulation of margins at each level:

The manufacturer might sell to a warehouse distributor (WD) at $18–22, representing their manufacturing cost plus margin.

The WD adds their margin and sells to a jobber at $28–35.

The jobber marks up to $40–50 for professional installers or $55–65 for retail customers.

If an installer charges $60 for the part (plus labor), they're making $10–20 on the part itself.

These numbers vary significantly by product category, quality tier, and competitive situation—but the structure is consistent. Each level of the distribution chain needs margin to survive; that accumulated margin determines the final price.

Where Value Gets Created—and Captured

Manufacturers capture value through production efficiency, technical differentiation, and brand strength. A manufacturer with lower production costs can offer better margins to distribution while maintaining competitive pricing. One with premium brand recognition can command higher prices.

Distributors capture value through logistics efficiency, service levels, and assortment breadth. A WD that can deliver any brake application overnight creates value that justifies their margin. One with a broader product line reduces the number of suppliers a jobber must manage.

Installers capture value through technical expertise, convenience, and trust. The installer who correctly diagnoses a brake problem, installs quality parts, and backs the work creates value worth paying for.

E-commerce disruptors capture value by compressing the traditional chain. By selling direct to consumers or installers, online retailers can offer lower prices while maintaining margins—though they sacrifice the service elements traditional distribution provides.

PRICING STRATEGY

Aftermarket pricing is both art and science. The right price depends on positioning, competition, and channel.

Price Positioning by Tier

The aftermarket spans distinct quality and price tiers, each serving different customer needs:

Premium/OE-equivalent parts command prices at or above OE dealer parts. Customers choosing this tier want the best, accept no compromises, and are less price-sensitive. They're often dealer service departments, performance enthusiasts, or customers with newer vehicles.

Standard aftermarket parts offer good quality at prices below OE. This is the largest segment—professional installers who want reliable parts without premium pricing, and value-conscious consumers who want quality without overpaying.

Economy parts compete primarily on price, sometimes at a fraction of OE pricing. Quality varies widely. Some economy parts perform acceptably; others are genuinely inferior. This tier serves extremely price-sensitive buyers and shops competing primarily on repair price.

Pricing must match positioning. Premium products priced too low lose credibility; economy products priced too high lose sales. Consistent positioning across the product line builds brand identity.

Program Pricing and Rebates

Large aftermarket buyers—major WDs, buying groups, national chains—negotiate program pricing that includes base prices plus rebate structures:

Volume rebates reward total purchase volume with end-of-period payments.

Growth rebates incentivize year-over-year purchase increases.

Marketing rebates compensate for co-op advertising, catalog listings, and promotional activities.

Prompt payment rebates (or penalties) address payment timing.

These rebates significantly affect real net cost. A program with attractive base pricing but thin rebates might actually be more expensive than one with higher list prices and generous rebates.

MAP Policies

MAP (Minimum Advertised Price) policies restrict how low prices can be publicly advertised. They protect brand positioning and prevent race-to-the-bottom pricing that damages margins throughout the channel.

MAP applies to advertised prices, not transaction prices. A retailer can sell below MAP; they just can't advertise that they're doing so. The distinction creates complexity—shoppers who add items to cart may see lower "in-cart" prices than advertised.

MAP enforcement requires commitment. Brands must monitor compliance and enforce consequences for violations. Inconsistent enforcement undermines the entire policy.

BRAND BUILDING

In the aftermarket, brand recognition drives sales. An installer reaching for a brake pad box is more likely to choose a name they know and trust. A consumer searching online is more likely to click on a recognized brand.

Building Professional Loyalty

Professional installers—the shops doing most brake work—choose parts based on:

Experience: What's worked well in the past? Which brands have caused callbacks? Installers remember problems more vividly than successes.

Technical support: Who helps when there's a problem? Brands that provide installation tips, torque specs, and diagnostic assistance earn loyalty.

Training: Brands that invest in installer training build relationships while improving installation quality. Training programs, certification, and technical seminars all contribute.

Warranty support: How easily are warranty claims processed? Brands that fight warranty claims damage relationships with the installers who influence purchasing decisions.

Coverage: Does the brand offer the applications installers need? Gaps in coverage send installers to competitors—who may then become the default choice.

Consumer Brand Awareness

Most consumers have no idea what brake pads are on their vehicles and don't particularly care. But a growing segment does research before repair visits, comparing options and reading reviews.

For these consumers, brand presence matters:

Online reviews on parts sites, Amazon, forums, and enthusiast communities shape perceptions. Brands must monitor and respond to reviews, addressing problems publicly when appropriate.

Content marketing—how-to videos, maintenance guides, technical explanations—builds awareness while providing genuine value.

Packaging and presentation affect perception. Premium brands need premium appearance. A well-designed box that looks professional on the installer's shelf reinforces quality positioning.

Enthusiast community engagement builds credibility with influential buyers. Sponsoring racing teams, supporting car clubs, and participating in enthusiast forums creates organic advocacy.

DISTRIBUTION STRATEGY

Getting products to market requires distribution strategy matching the brand's positioning and capabilities.

Traditional Distribution Relationships

Working through WDs and jobbers remains the primary path to market for most aftermarket suppliers.

Getting listed requires demonstrating value: competitive pricing, broad coverage, marketing support, and sales potential. Major WDs receive pitches constantly; standing out requires clear differentiation.

Maintaining shelf space requires performance. Products that don't sell get delisted. Marketing support, sales training, and promotional programs help keep products moving.

Exclusivity is a double-edged sword. Exclusive distribution arrangements can cement relationships and justify investment—but they also limit market access. Most brands work with multiple distributors, accepting the complexity for broader reach.

National Account Programs

Large retail chains (AutoZone, O'Reilly, Advance, NAPA) represent concentrated buying power. Landing a national account can transform a brand's volume—but comes with demanding requirements:

Pricing pressure: National accounts leverage their volume for aggressive pricing. Margins on this business are typically thinner than traditional distribution.

Inventory management: Suppliers may be required to hold consignment inventory, manage vendor-managed inventory (VMI) programs, or meet strict fill rate requirements.

Private label competition: Many national accounts push their own private label brands, which may directly compete with name-brand offerings.

Compliance requirements: EDI integration, specific packaging, labeling standards, and sophisticated logistics requirements add operational complexity.

E-Commerce and Direct Sales

Online channels offer opportunities and challenges:

Marketplace presence on Amazon, eBay, and parts-specific sites like RockAuto reaches consumers directly. But marketplace competition is intense, and platform fees compress margins.

Direct-to-consumer sales through brand websites offer better margins and customer relationship ownership. But building traffic and fulfillment capability requires investment.

Channel conflict is inevitable. Selling direct at lower prices than traditional distribution partners creates tension. Many brands maintain MAP pricing across all channels or deliberately price direct sales higher to protect distributor relationships.

COVERAGE STRATEGY

The aftermarket serves a vast array of vehicles—every make, model, and year currently on the road, plus classics and specialty applications. Coverage decisions significantly affect market access.

Understanding VIO

VIO (Vehicles in Operation) data tells you how many of each vehicle type are currently on the road. This data, available from providers like Experian and Polk, informs coverage priorities.

Not all vehicles matter equally. A part covering 5 million VIO has more sales potential than one covering 500,000. But competition matters too—commodity applications face more pricing pressure than specialized ones.

The coverage game involves balancing breadth against inventory investment. More SKUs mean more applications covered—but also more inventory carrying cost and complexity.

Consolidation and Supersession

Smart suppliers use part consolidation and supersession to maximize coverage with minimum SKUs:

Consolidation identifies where one part number can cover multiple applications. A single rotor design might fit several related vehicles.

Supersession chains link old part numbers to current ones as applications evolve, maintaining access to historical data while rationalizing active inventory.

ACES and PIES data standards enable electronic catalog systems that manage these relationships, matching parts to applications across the industry.

Managing the Long Tail

Popular applications account for the majority of volume, but the long tail of lower-volume applications adds up. Strategies for the long tail include:

Specialized distribution: Some WDs focus specifically on low-volume, hard-to-find applications, commanding premium pricing for their specialty expertise.

Made-to-order production: Rather than stock low-volume applications, some manufacturers produce to order, accepting longer lead times for reduced inventory risk.

Strategic gaps: Sometimes the right answer is not covering an application. If volume is too low and competition too intense, the inventory investment may not pay back.

E-COMMERCE DISRUPTION

Online retail has fundamentally changed aftermarket dynamics over the past decade, and the evolution continues.

What Changed

Price transparency increased dramatically. Consumers can instantly compare prices across dozens of sellers. Pricing that was once opaque became visible.

Access expanded. A consumer in a rural area with one local parts store can now choose from thousands of options online.

Information democratized. Forums, YouTube videos, and review sites give consumers unprecedented ability to research parts and evaluate options.

Winning Online

Success in e-commerce requires different capabilities:

Content quality matters enormously. Detailed product descriptions, fitment data, installation instructions, and quality images drive conversion. Poor content loses sales regardless of price.

Reviews and ratings heavily influence purchasing. Actively soliciting reviews and responding to negative feedback is essential.

Search optimization—both on marketplaces and through Google—drives visibility. Products that don't appear in search results don't sell.

Fulfillment speed has become table stakes. Amazon has trained consumers to expect two-day or faster delivery. Meeting this expectation requires inventory positioning and logistics infrastructure.

Managing Channel Conflict

Selling online while maintaining traditional distribution relationships creates inherent tension. Strategies for managing conflict include:

Price parity across channels—ensuring online prices don't undercut distribution partners.

Differentiated products—offering distinct SKUs or packages online versus through distribution.

MAP enforcement—maintaining consistent minimum advertised pricing everywhere.

Transparent communication—being honest with distribution partners about online strategy.

KEY TAKEAWAYS

- Aftermarket margins are better than OE (often several times higher) but competition is more fragmented and brand building is essential.
- The margin stack from manufacturer through distribution to installer means retail prices are typically 3–4x manufacturing cost.
- E-commerce is disrupting traditional distribution, creating both opportunity (direct customer access) and challenge (channel conflict with established partners).
- Success requires broad application coverage, strong distributor relationships, and brand recognition with both installers and consumers.

29. ELECTRONIC BRAKE SYSTEMS EXPLAINED

Modern brake systems increasingly depend on electronic controls to deliver safety and performance beyond what purely mechanical systems can achieve. This chapter bridges the gap between traditional brake engineering and the electronic systems that now govern braking.

FROM MECHANICAL TO ELECTRONIC

The evolution from purely mechanical to electronically-controlled braking happened gradually over decades, but the transformation is now essentially complete. Understanding this evolution provides context for current systems.

Anti-lock Braking Systems (ABS)

ABS was the first widespread application of electronic control to braking, and remains the foundation for more advanced systems.

The problem ABS solves: A locked (skidding) wheel provides less braking force than a wheel at the edge of lockup, and provides no steering control. ABS detects impending lockup and modulates brake pressure to keep wheels at optimal slip—maximizing braking force while maintaining steering control.

FIGURE 8: Electronic Brake System Evolution

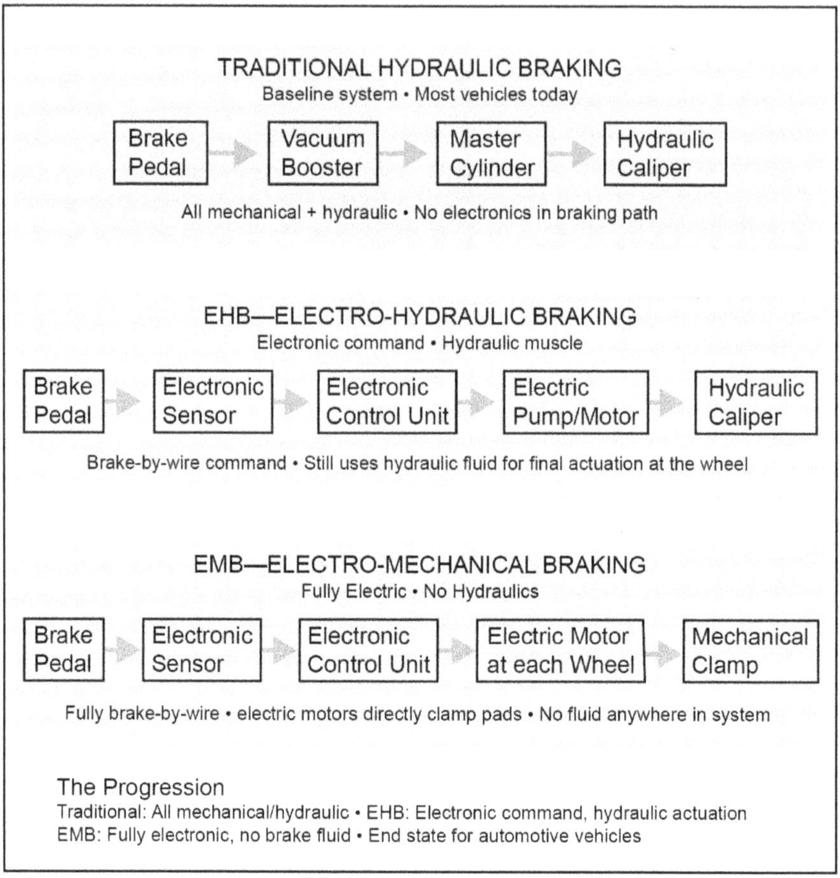

How ABS works: Wheel speed sensors at each wheel report rotation speed to an electronic control unit (ECU). The ECU compares individual wheel speeds and calculates vehicle reference speed. When a wheel's deceleration indicates impending lockup, the ECU commands the hydraulic modulator to reduce pressure to that wheel's brake. As the wheel recovers, pressure is gradually restored. This cycle repeats many times per second.

ABS components include:

Wheel speed sensors—typically variable reluctance or Hall effect sensors that produce signals proportional to wheel rotation speed.

Electronic control unit (ECU)—the computer that processes sensor inputs and makes control decisions.

Hydraulic modulator—an electro-hydraulic assembly containing solenoid valves and possibly a pump that can reduce, hold, and restore brake pressure to individual wheels.

Electronic Stability Control (ESC/ESP)

ESC builds on ABS by adding the ability to selectively apply brakes to correct vehicle dynamics, not just prevent wheel lockup.

The problem ESC solves: Vehicles can lose control when cornering forces exceed tire grip—either understeering (front slides, vehicle goes straight) or oversteering (rear slides, vehicle spins). ESC detects these conditions and intervenes.

How ESC works: In addition to wheel speed sensors, ESC uses a steering angle sensor (to know where the driver intends to go) and an inertial measurement unit (measuring yaw rate and lateral acceleration). The ECU compares driver intent with actual vehicle motion. When they diverge significantly, the system applies individual wheel brakes to generate corrective yaw moment—typically braking an outer front wheel to correct oversteer or an inner rear wheel to correct understeer.

ESC was proven so effective at reducing fatal crashes that it became mandatory on all new vehicles in the US (2012), EU (2014), and other markets.

Brake-by-Wire Evolution

The term "brake-by-wire" encompasses systems that replace mechanical or hydraulic connections between driver input and wheel brakes with electronic control. The evolution has proceeded in stages:

Electronically-controlled hydraulic brakes (early ABS and ESC) maintain conventional hydraulic connection from pedal to brakes but add electronic modulation capability.

Electro-hydraulic brakes (EHB) decouple the pedal from direct hydraulic connection. The driver's pedal input is measured by sensors; electronics determine desired braking and command actuators that generate hydraulic pressure. The pedal may connect to a simulator that provides artificial pedal feel.

Electro-mechanical brakes (EMB) eliminate hydraulic fluid entirely. Electric motors at each wheel directly clamp the brake caliper. This is "true" brake-by-wire—no fluid, no hydraulic connection whatsoever.

Each step increases electronic content and control authority while reducing mechanical complexity. The trade-off is that electronic failure becomes increasingly serious—the system must be designed to remain safe even with component failures.

SENSOR TECHNOLOGIES

Modern brake systems depend on sensors to understand vehicle and wheel dynamics.

Wheel Speed Sensors

Variable reluctance sensors use a toothed ring (tone ring) rotating past a magnetic sensor. As teeth pass the sensor, they create voltage pulses whose frequency indicates rotation speed. Simple and robust, but signals weaken at low speeds.

Active (Hall effect) sensors use semiconductors that respond to magnetic field changes. They provide cleaner signals, work better at low speeds, and can even indicate direction of rotation. Most new vehicles use active sensors.

Sensor mounting must be precise—air gap between sensor and tone ring is critical. Debris, bearing wear, or mounting issues can cause erratic signals that trigger ABS faults.

Inertial Sensors

Yaw rate sensors measure rotation around the vehicle's vertical axis—whether the vehicle is spinning or rotating as intended during cornering.

Lateral acceleration sensors measure side-to-side acceleration—how hard the vehicle is cornering.

These sensors typically use MEMS (micro-electromechanical systems) technology—tiny mechanical structures on silicon chips that respond to inertia. MEMS sensors are compact, reliable, and inexpensive enough for mass-market use.

Pedal Position and Pressure Sensors

In brake-by-wire systems, the driver's pedal input must be accurately sensed:

Pedal travel sensors measure how far the pedal has moved.

Pedal force sensors measure how hard the driver is pushing.

Brake pressure sensors measure actual hydraulic pressure in the system.

Redundancy is essential—multiple sensors measure the same input so the system can detect sensor failures. Disagreement between redundant sensors triggers fault handling.

ARCHITECTURE AND INTEGRATION

Modern brake systems don't exist in isolation—they're integrated with vehicle networks and other systems.

Network Communication

Brake ECUs communicate via vehicle networks, typically CAN (Controller Area Network) bus. This allows:

Information sharing—brake systems receive information from other systems (engine torque, transmission state, steering angle) and provide information to others (wheel speeds, brake status).

Coordinated control—brake and powertrain systems coordinate during regenerative braking, traction control, and stability interventions.

Diagnostic access—technicians can read fault codes, view live data, and command actuators through diagnostic tools connected to the network.

ADAS Integration

Advanced Driver Assistance Systems depend on brake system integration:

Automatic Emergency Braking (AEB) receives collision warnings from cameras and radar; the brake system must respond to brake commands generated by ADAS rather than the driver.

Adaptive Cruise Control maintains following distance by commanding both acceleration and deceleration—the brake system must accept and execute these commands.

Lane Keeping Assist may use gentle brake application on one side to provide steering correction.

Traffic Jam Assist combines these functions for semi-automated driving in congested traffic.

Regulations are lagging the technology in this space. One example: when an adaptive cruise control system slows a vehicle, should the brake lights activate? Many systems use a combination of throttle management and light braking—if the system simply reduces throttle to slow the vehicle, should the stop lamps illuminate? Questions like these still lack clear regulatory answers, and similar gaps exist across other ADAS-initiated braking scenarios.

Each integration point adds complexity and requires careful interface definition. The brake system must trust that commands from ADAS are legitimate while also maintaining the ability to override if something seems wrong.

FUNCTIONAL SAFETY

As electronic systems assume greater control over braking, ensuring their safety becomes paramount. The automotive industry addresses this through systematic functional safety engineering.

ISO 26262

ISO 26262 is the functional safety standard for automotive systems. It defines:

Hazard analysis processes to identify what can go wrong and how serious the consequences are.

Automotive Safety Integrity Levels (ASILs) that classify hazard severity from ASIL A (lowest) to ASIL D (highest). Brake systems typically fall into ASIL C or D—the most stringent levels.

Requirements for hardware and software design, verification, and validation at each ASIL level.

Brake systems must be designed, developed, and tested according to ISO 26262 requirements. Documentation must demonstrate systematic analysis of potential failures and mitigations.

Redundancy and Fault Tolerance

High-integrity brake systems incorporate multiple layers of protection:

Sensor redundancy—multiple sensors measure critical inputs; the system can detect disagreement and fall back to safe behavior.

Actuator redundancy—multiple actuators may provide braking; failure of one leaves others functional.

Power supply redundancy—backup power (second battery or supercapacitors) ensures braking remains available if the primary electrical system fails.

Communication redundancy—critical networks may be duplicated.

The goal is fail-operational design where the system continues to function (perhaps with degraded performance) even after failures, rather than simply failing safely.

CYBERSECURITY CONSIDERATIONS

As brake systems become networked and software-defined, cybersecurity becomes a safety concern.

Attack Vectors

Potential attack paths include:

Diagnostic ports that provide deep system access may be exploited if physical access is obtained.

Wireless interfaces (Bluetooth, WiFi, cellular) create remote attack possibilities.

Third-party connected devices (aftermarket telematics, OBD dongles) may bridge from less-secure networks to vehicle systems.

Supply chain attacks could compromise software or hardware before it reaches the vehicle.

Protections

Automotive cybersecurity standards (ISO/SAE 21434) define requirements for:

Threat assessment to identify potential attacks and their impacts.

Security architecture separating critical systems from less-trusted networks.

Secure boot ensuring only authenticated software runs on ECUs.

Encryption and authentication of network messages.

Intrusion detection to identify anomalous behavior.

For brake systems specifically, the priority is ensuring that safety-critical functions cannot be compromised by network attacks. This typically means gateway architectures that limit what messages can reach brake ECUs from external interfaces.

KEY TAKEAWAYS

- ABS prevents wheel lockup; ESC adds yaw control to prevent loss of vehicle control—ESC alone reduces fatal crashes significantly and is now mandated.
- Brake-by-wire systems (EHB and EMB) replace mechanical connections with electronic control, enabling integration with ADAS and autonomous systems.
- Functional safety standards (ISO 26262) require systematic hazard analysis, redundancy, and rigorous validation for brake electronics.
- Cybersecurity is an emerging concern as networked brake systems create potential attack vectors—protection requires secure architecture and ongoing vigilance.

CONCLUSION: THE ROAD AHEAD

After years of covering this industry and working with the people who lead it, I'm more convinced than ever that brakes deserve more attention than they receive.

They're the ultimate safety system—the technology that stands between normal driving and disaster. They embody some of the most demanding engineering challenges in the automotive world: extreme temperatures, precise tolerances, absolute reliability. And they're evolving rapidly in response to electrification, autonomy, and environmental pressures.

Let me share what I see looking forward:

Consolidation will continue. The capital requirements for electronic braking systems development, combined with the scale needed to serve global automakers, will drive further industry consolidation. Smaller players will be acquired or pushed to the margins. The competitive landscape in 2030 will look different than today.

Software will define value. The companies that thrive will be those that master software-defined braking—not just the hardware, but the algorithms, the integration, the data. Traditional brake companies must become software companies, or partner with those who are.

Regulation will accelerate innovation. Euro 7 and similar standards will force technology advancement, particularly in low-emission braking. Companies that develop compliant solutions first will gain competitive advantage. Those that lag may find themselves locked out of major markets.

New entrants will challenge incumbents. The transition to EVs and electronic braking systems opens opportunities for companies not burdened by legacy hydraulic system investments. Chinese suppliers are moving fast. Tech-adjacent companies may enter the space. Disruption is possible.

Brakes will become invisible—until needed. In the autonomous future, most people will never think about brakes. The systems will work seamlessly, managed by software, activated autonomously when needed. But for the engineers and manufacturers who create those systems, the responsibility will be greater than ever. Getting it wrong isn't an option.

This industry stands at an inflection point. The next five years will determine which companies lead the transition to software-defined braking—and which get left behind. The winners will be those who invest in electronic integration, navigate regulatory compliance, and build the talent needed to compete in an increasingly technical landscape.

For those of us in this industry, the opportunity is significant. Braking is moving from commoditized hardware toward high-value integrated systems. The companies and professionals who position themselves at that intersection—where mechanical expertise meets software capability—will capture disproportionate value.

I hope this handbook serves as a useful reference as you navigate the changes ahead. The industry will continue to evolve, and I'll continue covering it through *The BRAKE Report* and placing its leaders through Hagman Search.

If you have feedback, insights to share, or just want to connect—reach out. This industry has always been built on relationships.

—Brian Hagman

ABOUT THE AUTHOR

Brian Hagman is the founder and publisher of *The BRAKE Report*, the leading trade publication serving the global brake industry, and the founder of Hagman Search, an executive recruiting firm focused exclusively on the brake sector.

Since launching *The BRAKE Report* in 2012, Brian has established it as the go-to resource for brake industry professionals seeking specialized coverage. Through the website, newsletter, podcast, and social media platforms, *The BRAKE Report* reaches brake industry professionals in over 100 countries—from C-level executives at major manufacturers to engineers, product managers, sales leaders, and supply chain professionals across the brake ecosystem.

Through Hagman Search, Brian leverages his deep industry network and expertise to place executives and technical professionals at brake companies worldwide. His dual perspective—seeing the industry through both its public stories and its confidential leadership decisions—informs the unique insights in this handbook.

Brian also publishes additional automotive trade media through Hagman Media, covering the electric vehicle and autonomous driving sectors that are reshaping the automotive industry and its approach to braking technology.

Connect

Phone: +1 270 300 7001
Email: bhagman@hagmanmedia.com | bhagman@hagmansearch.com
LinkedIn: linkedin.com/in/brianhagman

Websites

Hagman Search: www.hagmansearch.com
Hagman Media: www.hagmanmedia.com
The BRAKE Report: www.thebrakereport.com

APPENDIX A: GLOSSARY OF TERMS

ABS (Anti-lock Braking System): Electronic system that prevents wheel lockup during hard braking by rapidly modulating brake pressure, allowing the driver to maintain steering control.

ACES (Aftermarket Catalog Exchange Standard): Data format that defines how aftermarket parts application data is structured—which parts fit which vehicles. Enables electronic catalogs used by distributors, jobbers, and online retailers.

ADAS (Advanced Driver Assistance Systems): Electronic systems that assist drivers, including features like automatic emergency braking, adaptive cruise control, and lane-keeping assistance.

AEB (Automatic Emergency Braking): System that automatically applies brakes when sensors detect an imminent collision and the driver has not responded.

Aftermarket: Replacement parts and accessories sold after the original vehicle purchase, as opposed to original equipment.

AIAG (Automotive Industry Action Group): Organization that publishes the core automotive quality tool manuals, including APQP, PPAP, FMEA, SPC, and MSA.

AK Master (SAE J2522): Comprehensive friction material evaluation procedure widely used in Europe and globally, consisting of multiple test sequences that assess fade, recovery, speed sensitivity, and other performance characteristics.

APQP (Advanced Product Quality Planning): Structured process for developing automotive products, defining phases and deliverables from concept through production launch.

ASIL (Automotive Safety Integrity Level): Risk classification from ISO 26262 ranging from ASIL A (lowest) to ASIL D (highest). Brake systems typically require ASIL C or D.

Backing Plate: Metal plate to which friction material is bonded to create a brake pad.

Brake Fade: Reduction in braking effectiveness due to overheating, caused by the friction material's coefficient of friction decreasing at elevated temperatures.

Brake Fluid: Hydraulic fluid used to transmit force from the brake pedal to the wheel brakes. Common types include DOT 3, DOT 4, and DOT 5.1.

Brake-by-Wire: Braking systems that use electronic signals rather than mechanical or hydraulic connections to transmit the driver's braking command.

Caliper: Component that houses the brake pads and pistons, clamping the pads against the rotor when hydraulic pressure is applied.

CAN Bus (Controller Area Network): Standard vehicle communication network that allows electronic control units to share data without a central computer.

Carbon-Ceramic: Composite brake disc material made from carbon fiber and ceramic, offering exceptional heat resistance and weight savings.

Coefficient of Friction (CoF): Measure of the friction between two surfaces; in braking, the friction between pad and rotor that determines stopping power.

Copper-Free: Friction formulations containing less than 0.5% copper by weight, mandated by legislation in Washington and California to reduce environmental contamination.

Cpk (Process Capability Index): Statistical measure of how well a process stays within specification limits. Customers typically require Cpk of 1.33 or higher for critical dimensions; 1.67 for safety-critical characteristics.

CVSA (Commercial Vehicle Safety Alliance): Organization coordinating commercial vehicle safety enforcement across North American jurisdictions.

DFMEA (Design Failure Mode and Effects Analysis): Risk assessment methodology applied to product design, identifying potential failure modes, their severity, likelihood, and detectability—used to drive design improvements before production.

Disc Brake: Brake design where pads clamp against a spinning disc (rotor) to generate friction.

DOT 3, DOT 4, DOT 5, DOT 5.1: U.S. Department of Transportation brake fluid classifications. DOT 3, 4, and 5.1 are glycol-based and hygroscopic (absorb moisture); DOT 5 is silicone-based and not compatible with ABS systems. Higher numbers generally indicate higher boiling points and better high-temperature performance.

Drum Brake: Brake design where shoes press outward against the inner surface of a spinning drum.

DTV (Disc Thickness Variation): Measurable variation in rotor thickness around its circumference, a primary cause of brake judder and pedal pulsation.

EBD (Electronic Brake-force Distribution): System that automatically varies brake force between front and rear wheels based on load and driving conditions.

EBS (Electronic Braking System): Electronic control of air brake systems in commercial vehicles.

eBooster (Bosch): Bosch's second-generation electro-hydraulic brake actuation system for hybrid and electric vehicles. Provides electronic control of braking while retaining hydraulic backup, enabling seamless regenerative braking integration.

ECU (Electronic Control Unit): Computer that controls electronic vehicle systems, including braking functions.

Edge Code: Two-letter friction coefficient designation (one for normal conditions, one for hot conditions) stamped on the edge of every brake friction material sold in the United States. Required for AMECA certification and standardized through the Chase Test (SAE J661).

8D (Eight Disciplines): Structured problem-solving methodology used to respond to quality issues. Requires teams, containment, root cause analysis, corrective actions, verification, prevention, and recognition.

EHB (Electro-Hydraulic Brake): Brake-by-wire system that uses electronic control with hydraulic actuation at the wheels.

EMB (Electro-Mechanical Brake): Brake-by-wire system that uses electric motors at each wheel to actuate the brakes, eliminating hydraulic fluid entirely.

ESC (Electronic Stability Control): System that helps prevent loss of control by automatically applying brakes to individual wheels.

Euro 7: European Union emission standard that for the first time includes limits on brake particulate emissions (PM10), driving changes in friction material formulation.

Fade: See Brake Fade.

FAST Machine (Friction Assessment Sample Test): Test device developed by Dr. Arne Anderson at Ford Research in the mid-1960s that evaluates a half-square-inch friction material sample through a 90-minute constant-force, constant-speed procedure. Widely used for formulation development and quality control.

Fixed Caliper: Caliper design with pistons on both sides of the rotor, rigidly mounted. Offers superior stiffness and pedal feel, used primarily on performance and luxury vehicles.

Floating Caliper: Caliper design that slides on pins, using pistons on only one side to clamp both pads against the rotor.

FMCSA (Federal Motor Carrier Safety Administration): U.S. agency regulating commercial motor vehicle safety, including brake inspection and maintenance requirements.

FMEA (Failure Mode and Effects Analysis): Risk assessment methodology that systematically evaluates potential failure modes, their severity, likelihood, and detectability. Applied to both product design (DFMEA) and manufacturing processes (PFMEA).

FMSI (Friction Materials Standards Institute): North American organization that assigns standardized D-numbers to brake pad and shoe applications, creating the common language used across aftermarket catalogs and distribution systems.

FMVSS (Federal Motor Vehicle Safety Standards): U.S. regulatory standards for vehicle safety, including braking performance requirements.

Foundation Brake: The basic mechanical braking components at each wheel (caliper, pads, rotor, or drum and shoes).

Friction Material: The compound that creates friction when pressed against the rotor or drum; the "business end" of the brake pad or shoe.

Gladhand: Standardized air coupling connecting tractor and trailer air systems, using color-coded connectors: red for supply line, blue for service line.

IATF 16949: The automotive-specific quality management standard built on ISO 9001. Required for all serious automotive suppliers; losing certification effectively means losing the ability to supply OE customers.

iBooster (Bosch): Bosch's electro-hydraulic brake actuation system that replaces the conventional vacuum brake booster with an electric motor-driven unit. Widely adopted for EV and hybrid applications because it enables precise regenerative braking blending without requiring engine vacuum.

ISO 26262: International standard for functional safety of automotive electrical and electronic systems, defining hazard analysis processes and safety integrity levels.

Job 1: Industry term for the start of series production of a new vehicle program, a critical milestone for both OEMs and suppliers.

Judder: Vibration felt through the brake pedal or steering wheel during braking, often caused by rotor thickness variation.

Keiretsu: Japanese supplier network model characterized by vertical integration, long-term relationships, partial equity ownership, and deep technical collaboration. Makes it exceptionally difficult for outside suppliers to win programs at Japanese OEMs.

Low-steel: Friction material formulation combining an organic base with small amounts of steel for improved heat handling. Dominant in European applications. Sometimes called low-metallic.

Master Cylinder: Hydraulic pump that converts pedal force into hydraulic pressure distributed to the wheel brakes.

MEMS (Micro-Electromechanical Systems): Miniaturized mechanical structures on silicon chips used in automotive sensors, including yaw rate and acceleration sensors in brake systems.

MSA (Measurement System Analysis): Statistical evaluation of gauges and measurement methods to ensure they are reliable enough to detect the variation being controlled. A required element of PPAP.

NAO (Non-Asbestos Organic): Friction material formulation using organic compounds without asbestos or significant metal content.

NVH (Noise, Vibration, Harshness): Engineering discipline focused on reducing unwanted noise, vibration, and harshness in vehicles, including brake noise.

OE/OEM (Original Equipment/Manufacturer): Parts or suppliers that provide components for new vehicle production.

OES (Original Equipment Service): Replacement parts sold through vehicle dealer service departments, typically sourced from the same suppliers who provided OE production parts.

Pad: Brake pad; the friction material assembly that presses against the rotor in a disc brake system.

Pedal Feel: The sensation transmitted through the brake pedal to the driver, including travel, effort, and feedback.

Pedal Simulator: Device in electro-hydraulic brake systems that provides artificial pedal feel to the driver when the pedal is decoupled from direct hydraulic connection.

PFMEA (Process Failure Mode and Effects Analysis): Risk assessment methodology applied to manufacturing processes, identifying where process variation could produce defective parts and driving process control improvements.

Phenolic Resin: Thermoset polymer commonly used as the binder in friction material formulations, holding other ingredients together under extreme heat and pressure.

Piece Price: The per-unit cost paid for each component delivered in OE supply, distinct from tooling costs, engineering recovery, and other program charges.

PIES (Product Information Exchange Standard): Data format that defines aftermarket product attributes such as dimensions, features, images, and marketing descriptions. Complements ACES in powering electronic cataloging.

PM10: Particulate matter with diameter of 10 micrometers or less; regulated in Euro 7 brake emissions standards.

PPAP (Production Part Approval Process): Documented evidence that a production process can consistently produce parts meeting all engineering design and specification requirements.

PPM (Parts Per Million): Universal quality metric in automotive manufacturing. A target of 10 PPM means no more than 10 defective parts per million delivered.

Proportioning Valve: Device that adjusts brake pressure distribution between front and rear wheels to account for weight transfer during braking.

Regenerative Braking: System that captures kinetic energy during deceleration, converting it to electrical energy stored in a battery.

Relay Valve: Air brake component that speeds brake response by using a small control signal to trigger release of locally stored air pressure.

RFI (Request for Information): Exploratory inquiry from an OEM to potential suppliers, gathering information about capabilities and technology before formal sourcing.

RFQ (Request for Quotation): Formal bid solicitation from an OEM with detailed specifications and commercial requirements, requesting technical proposals and pricing.

Rotor: The spinning disc in a disc brake system against which the pads create friction; also called a brake disc.

RPN (Risk Priority Number): Score used in FMEA calculated by multiplying severity, occurrence, and detection ratings. Higher RPNs drive prioritization of corrective actions.

RSC (Roll Stability Control): Commercial vehicle safety system that detects impending rollover using body roll sensors and automatically applies brakes to reduce speed.

SAE J2521: Industry-standard dynamometer test for brake noise evaluation. Subjects brake assemblies to a series of stops under varying conditions to assess squeal, groan, and other noise issues.

S-cam: S-shaped cam mechanism in commercial vehicle drum brakes that rotates to push brake shoes outward against the drum when air pressure is applied.

Scorecard: Systematic OEM measurement of supplier performance across quality, delivery, cost, and technology dimensions that directly influences future sourcing decisions.

Semi-Metallic: Friction material formulation containing significant metal content for heat dissipation and stopping power.

Sensify (Brembo): Brembo's software-defined brake-by-wire system that modulates braking force at each wheel independently in real time, based on vehicle dynamics and driving conditions. Enables over-the-air tuning of brake feel and performance.

Slack Adjuster: Air brake component connecting the chamber pushrod to the S-cam shaft, converting linear motion to rotary motion and compensating for shoe wear.

Spring Brake: Fail-safe brake on commercial vehicles where a powerful spring applies the brake and air pressure releases it. Loss of air automatically stops the vehicle.

SPC (Statistical Process Control): Use of statistical methods to monitor and control manufacturing processes, detecting drift or out-of-specification trends before they produce defective parts.

SQE (Supplier Quality Engineer): The customer's representative responsible for evaluating and approving a supplier's quality performance. SQEs have effective veto power over production decisions, including PPAP approval.

Squeal: High-frequency brake noise caused by vibration of brake components.

Tier 1: Direct supplier to vehicle manufacturers (OEMs).

Tier 2: Supplier to Tier 1 companies rather than directly to vehicle manufacturers.

Tier 3: Supplier to Tier 2 companies.

Tone Ring: Toothed ring mounted on a rotating component that generates signals for wheel speed sensors as teeth pass the sensor.

Two-Piece Rotor: Brake disc design separating the friction ring from the center hat, allowing different materials for each and accommodating thermal expansion.

Unsprung Weight: Mass not supported by the vehicle suspension system, including wheels, tires, and brake components. Reducing it improves ride quality and handling.

Vented Rotor: Brake disc with internal vanes that promote airflow for improved cooling.

VIO (Vehicles in Operation): Data on how many of each vehicle make, model, and year are currently on the road. Used by aftermarket manufacturers to prioritize application coverage based on market opportunity.

WD (Warehouse Distributor): Aftermarket distribution company that stocks and distributes parts to jobbers and retailers.

APPENDIX B: MAJOR COMPANY DIRECTORY

This directory includes major brake industry companies organized by category and alphabetized within each section.

Profiles for the industry's largest and most widely encountered companies include additional market context. All companies listed are active participants in the global brake industry. Inclusion does not imply endorsement, and omission does not imply insignificance—the brake industry includes hundreds of companies worldwide, and this directory focuses on those most likely to be encountered by professionals navigating the industry.

BRAKE SYSTEM & COMPONENT MANUFACTURERS

ADVICS—Japan—Integrated brake systems, brake pads, and calipers. An Aisin subsidiary within the Toyota Group, ADVICS is one of the largest global brake suppliers by volume, serving as a primary brake system integrator for Toyota and Lexus vehicles worldwide.

Aisin Corporation—Japan—Integrated brake systems and chassis components, closely aligned with Toyota. One of the world's largest automotive suppliers by revenue, with braking operations conducted primarily through its ADVICS subsidiary.

Akebono Brake Industry—Japan—Friction materials and calipers with particular strength in ceramic and copper-free formulations. Pioneered low-noise ceramic pad technology and supplies both OE and aftermarket channels, with a significant manufacturing and engineering presence in North America.

Alcon Components—UK—High-performance brake and clutch components produced in low volumes for motorsports, military, and specialty vehicle applications. A respected British manufacturer whose products are found on racing cars across multiple series and on specialist vehicles where standard production components are inadequate.

Allied Nippon—India—Disc brake pads, brake linings, and brake shoes for motorcycles, passenger vehicles, and commercial vehicles. Serves the Indian domestic market and exports to aftermarket distributors across the Middle East, Africa, and Southeast Asia.

American Friction Technologies—USA—Sintered metallic brake pads for power sports, off-highway, and specialty vehicle applications. One of the few North American manufacturers focused exclusively on sintered friction technology for non-automotive segments.

ASIMCO Technologies—China—Global independent brake pad manufacturer supplying friction materials to both OE and aftermarket channels worldwide. One of China's earliest brake friction exporters to Western aftermarket distributors, with a customer base spanning over 60 countries.

ASK Automotive—India—Brake panel assemblies, brake shoes, and disc brake pads, primarily for two-wheeler applications. A key supplier to major Indian two-wheeler OEMs and increasingly active in export markets as Indian motorcycle manufacturers expand globally.

Astemo—Japan—Brake systems, ABS, and ESC units. Formed in 2021 from the merger of Hitachi Automotive Systems with Honda-affiliated suppliers Showa, Keihin, and Nissin, serving as a primary brake system supplier within the Honda ecosystem.

AUMOVIO SE—Germany—Electronic braking systems, brake-by-wire technology, and ADAS integration, with the ATE aftermarket brand. Spun off from Continental AG in September 2025 as an independent publicly traded company. AUMOVIO reports annual sales of roughly €19.4 billion and approximately 92,000 employees.

Continental AG is narrowing to a pure-play tires business, with the planned divestiture of its ContiTech group sector expected to complete in 2026.

Ausco Products—USA—OEM brakes for construction, agricultural, mining, turf care, and ATV/utility vehicle markets. A niche player focused on off-highway and industrial applications where duty cycles and environmental conditions differ significantly from on-road braking.

AUTOFREN SEINSA—Spain—Brake caliper repair kits, seals, boots, and rubber components for automotive spare parts and caliper rebuilders. A European specialist whose products are essential to the caliper remanufacturing industry across Europe and Latin America.

AYD Automotive Industry—Turkey—Brake discs, brake drums, brake pads, and a comprehensive range of steering and suspension components for passenger and light commercial vehicles. One of Turkey's largest automotive component companies, with over 4,500 employees and annual production exceeding 50 million units.

Baer Brakes—USA—Performance brake kits for the aftermarket, specializing in big brake upgrades for American muscle cars, trucks, and SUVs. A recognized name in the enthusiast market for bolt-on disc brake conversions and performance upgrades.

Bendix Commercial Vehicle Systems—USA—Air brakes, ABS/ESC, and advanced safety systems for commercial vehicles. A Knorr-Bremse subsidiary and the dominant supplier in North American heavy-duty truck and trailer braking, with a broad portfolio spanning foundation brakes, air treatment, and driver assistance technologies.

Beşer Balatacılık—Turkey—Brake pads, brake linings, and disc brake pads for passenger cars, trucks, trailers, buses, and commercial vehicles. Founded in 1969 in İzmir, Beşer is the largest brake friction manufacturer in Turkey, with four production sites, over 700 employees, and annual capacity of 8,000 tons of brake linings and 24 million disc brake pad sets.

Bethel Automotive (WBTL)—China—Chassis systems including EPB (electronic parking brake), WCBS (wire-controlled brake systems),

ADAS, and lightweight aluminum components. One of China's most prominent domestic brake-by-wire developers, supplying several major Chinese EV manufacturers.

Bosch (Robert Bosch GmbH)—Germany—Complete brake systems, ABS/ESC, and brake-by-wire technology. The world's largest automotive supplier, whose braking division benefits from deep synergies with their electronics, sensors, and software businesses. Their iBooster and eBooster 2.0 systems are widely adopted for EV regenerative braking integration.

BPW Bergische Achsen—Germany—Trailer axles, brakes, and running gear systems. A major European supplier of trailer and commercial vehicle chassis components, with strong market position in European trailer brake systems including disc and drum configurations.

Brakes India—India—Part of the TSF Group. One of the largest brake manufacturers in Asia, producing foundation brakes, calipers, drums, and rotors for both the Indian domestic market and global export. A significant supplier of ferrous castings for passenger and commercial vehicle applications worldwide.

Brembo S.p.A.—Italy—High-performance calipers, rotors, and complete braking systems with deep motorsports heritage. The most recognized premium brake brand globally, Brembo owns AP Racing, BYBRE, J.Juan, and SBS Friction, and has expanded aggressively into software-defined braking with its Sensify system.

Bremsen Technik—UK—Friction materials and related components offering bespoke solutions for commercial vehicles, passenger cars, and racing. A specialist formulator and manufacturer serving customers who require custom friction development rather than off-the-shelf products.

Bremskerl CV—USA—Commercial vehicle friction and braking products company focused exclusively on the North American commercial vehicle market. The company offers a growing portfolio of air disc brake pads, USA-remanufactured calipers, rotors, and related hardware for air disc and hydraulic applications.

Breyden—Germany—Brake discs, calipers, carriers, bridges, and backing plates. A vertically integrated European component manufacturer serving OE and Tier 1 customers who require precision-machined brake hardware.

BWI Group—USA—OE brake and suspension components for the automotive industry. Originally the chassis division of Delphi, BWI operates as a global chassis systems supplier with engineering centers in the U.S., Europe, and China.

Carlisle Brake & Friction—USA—High-performance and severe-duty brake, clutch, and transmission friction products for mining, construction, military, agricultural, and aerospace markets. Part of CentroMotion, the company also markets the Hawk Performance brand of street and motorsports brake pads, one of the most recognized names in the performance aftermarket. A specialist in extreme-duty applications where standard automotive friction technology cannot withstand the loads, temperatures, and environmental conditions involved.

Consolidated Metco (ConMet)—USA—Wheel ends, cast metal components, and aerodynamic devices for OE and aftermarket commercial vehicle segments. A leading supplier of wheel hubs and brake drums to major North American truck OEMs, with growing capabilities in lightweight and preset hub assemblies.

Cummins Drivetrain and Braking Systems (formerly Meritor)—USA—Commercial vehicle braking systems including axles, brakes, and drivetrain components. Meritor was acquired by Cummins Inc. in 2022 and operated as Cummins-Meritor until rebranded as Cummins Drivetrain and Braking Systems in 2025, now a business unit within Cummins' Components segment. The Meritor name is retained as the aftermarket brand, offering decades of established foundation brake expertise—particularly in drum and disc brakes for heavy trucks—under a broader powertrain and drivetrain umbrella.

Danaher—China—Brake pads, discs, drums, shoes, calipers, sensors, and complete brake kits. A large-scale Chinese manufacturer producing a comprehensive range of brake components primarily for export to aftermarket distributors in Europe, the Middle East, and the Americas.

Disc Brakes Australia (DBA)—Australia—Disc brake rotors, brake pads, drums, and complete braking systems. The leading Australian brake manufacturer, known for their patented slotted rotor designs and strong presence in the Australian, New Zealand, and North American performance aftermarket.

Dynamic Friction Company (DFC)—USA—Aftermarket brake components including pads, rotors, and brake kits. A growing presence in the value-oriented aftermarket segment, competing on broad application coverage and aggressive pricing through both traditional distribution and e-commerce channels.

DuraBrake—USA—Brake drums, lightweight composite brake drums, air disc brake rotors, hubs, pads, and calipers for trucks, trailers, and buses. Focused on the North American heavy-duty segment with particular emphasis on weight-saving solutions that help fleets maximize payload capacity.

EBC Brakes—UK—Performance brake pads and rotors. A strong brand among automotive enthusiasts worldwide, manufacturing in England with a product line spanning from premium street pads to track-oriented compounds, sold primarily through enthusiast retailers and e-commerce.

EGE Fren—Turkey—Brake assemblies and automotive components for commercial vehicle and industrial markets worldwide. One of Turkey's established brake manufacturers, leveraging the country's growing position as a supplier base for European commercial vehicle OEMs.

EKU—Turkey—Brake drums, brake discs, and brake pads for commercial and passenger vehicles. Founded in 1962 as Türkiye's first brake drum manufacturer, EKU has grown into Europe's largest foundry in its field, with a melting capacity of 350,000 tons and an annual machining capacity of 3.5 million units.

FDP Friction Science—USA—Disc pads and brake shoes supplied to OEMs, retailers, warehouse distributors, and brake manufacturers.

A friction specialist that also supplies semi-finished products to other brake brands for private-label programs.

Frasle Mobility—Brazil—Brake pads and linings for commercial and passenger vehicles, marketed under the Abtex, Fras-le, Fremax, Juratek, Jurid, and Fritec brands. The largest friction material manufacturer in Latin America, with a global portfolio built through strategic acquisitions that has expanded their reach well beyond the South American market.

Friction One—China—Premium braking products for retailers, distributors, and OEMs. Sister company includes the Magma Brakes brand. Positioned as a quality-focused Chinese manufacturer targeting the mid-to-premium aftermarket tiers in North America and Europe.

GALFER (Industrias Galfer)—Spain—Friction materials for both motorcycle and automobile industries. A Spanish manufacturer with particular strength in motorcycle brake pads and discs, supplying OE programs for European motorcycle manufacturers and a growing powersports aftermarket.

GMP Friction Products—USA—Sintered metallic brake pads, high-performance clutch components, and wet disc brakes. Serves industrial, off-highway, and specialty applications where sintered metallic friction technology is required for extreme duty cycles.

Grantex—Greece—Brake pads and linings for heavy-duty, passenger car, and industrial segments. A Southern European manufacturer with distribution networks across the Mediterranean, Middle East, and North African aftermarkets.

Haldex—Sweden—Brake adjusters, air suspension, and commercial vehicle components. Acquired by SAF-Holland in 2022. A global leader in automatic slack adjusters for trucks and trailers, with products found throughout the worldwide commercial vehicle market.

Hayes Performance Systems—USA—Brakes, suspension, and wheels for powersport, mountain bike, motorcycle, turf care, agriculture, construction, and defense applications. A diversified braking specialist whose hydraulic disc brake expertise spans from competitive mountain biking to military vehicles.

Hella—Germany—Aftermarket brake components including pads, discs, shoes, drums, calipers, and hoses. Part of the Forvia group, Hella leverages its strong European aftermarket distribution network to offer a comprehensive brake parts program alongside its core lighting and electronics business.

HL Mando—South Korea—Brake systems, calipers, and electronic controls. The primary brake system supplier to Hyundai-Kia, with a growing global OE customer base and significant investment in brake-by-wire and ADAS-integrated braking technology.

Hyundai Mobis—South Korea—Integrated brake systems and chassis modules. A captive supplier to Hyundai Motor Group and one of South Korea's largest automotive suppliers, providing complete brake system integration for Hyundai, Kia, and Genesis vehicles globally.

ICER Brakes—Spain—Brake pads, discs, shoes, linings, and wear indicators for automotive, commercial vehicles, and buses. A well-established Spanish friction manufacturer with strong European distribution and growing presence in Latin American markets.

Inroble—Canada—Brake pads, rotors, drums, and calipers for light and heavy duty applications. A Canadian manufacturer and distributor serving the North American aftermarket with products spanning both passenger vehicle and commercial vehicle segments.

ITT Motion Technologies—Italy—Brake pads for OE and aftermarket applications. Supplies friction materials to European OEMs and the global aftermarket, with particular expertise in low-emission and low-noise formulations that position them well for Euro 7 compliance.

KBAutosys—South Korea—Brake pads for the automotive segment. A Korean friction specialist serving domestic OEMs and the Korean aftermarket, with growing export business as Hyundai-Kia's global production footprint expands.

Knorr-Bremse—Germany—The global leader in commercial vehicle and rail braking systems. Dominates heavy-duty braking worldwide with comprehensive air brake system portfolios, and owns Bendix in North America. Expanding into driver assistance and electronic controls for commercial vehicle safety.

Lumag—Poland—Brake pads and linings for passenger cars and commercial vehicles, marketed under the Breck aftermarket brand. One of the largest friction material manufacturers in Eastern Europe, with Breck positioned as a value-quality brand across European aftermarket channels.

Marathon Brake Systems—USA—Brake linings and disc brake products for trucking and transit bus applications. A specialist commercial vehicle friction manufacturer serving North American fleets, transit authorities, and heavy-duty aftermarket distributors.

Master Power Brakes—USA—Model-specific disc brake conversion kits for classic cars, muscle cars, and hot rods. Serves the restoration and custom vehicle market with engineered solutions that replace original drum brake systems with modern disc brake assemblies.

Masu Brakes—India—Brake pads and brake assemblies for domestic OE and aftermarket channels. A growing Indian friction manufacturer that has expanded from two-wheeler components into passenger vehicle applications, with increasing exports to global aftermarket distributors.

MAT Holdings (GRI Engineering)—USA—Brake pads and rotors for multiple market segments. Supplies private-label brake products across multiple retail channels and maintains a diversified industrial portfolio that provides stability beyond the automotive brake business.

Max Advanced Brakes—Canada—Brake pads, rotors, and accessories manufactured to OE specifications. A Canadian aftermarket supplier positioning products at the premium tier with emphasis on OE-equivalent quality and comprehensive application coverage for the North American market.

Mevotech—Canada—Aftermarket chassis, steering, suspension, brakes, and driveline parts. A broad-line Canadian aftermarket supplier that has built a strong position in the professional installer channel through its TTX and Supreme lines of chassis and brake components.

Meyle—Germany—Aftermarket parts for cars, vans, and trucks sold in over 120 countries. A major German aftermarket brand offering brake components as part of a comprehensive vehicle parts program, with particular strength in the European independent aftermarket.

Momentum USA—USA—Aftermarket brake pads and rotors under the AmeriBRAKES brand, 100% North American manufactured. One of the few aftermarket brake suppliers that manufactures entirely in North America, positioning domestic production as a quality and supply chain reliability differentiator.

Neotek Corporation—Taiwan—Brake rotors and drums produced in fully vertically integrated factories. One of Taiwan's leading brake casting operations, manufacturing from raw iron through finished machined product, primarily for export to aftermarket distributors in North America and Europe.

NewTek Automotive—USA—Brake pads, rotors, drums, and shoes. Acquired China Yantai Friction Company (CYFC), combining U.S.-based product management and distribution with Chinese manufacturing scale to serve the North American value aftermarket.

Nisshinbo—Japan—Brake pads for automotive OE and aftermarket applications. One of Japan's major friction material manufacturers, supplying global OE programs and aftermarket channels with formulations developed for Japanese and international vehicle platforms.

NRS Brakes—Canada—Galvanized brake pads for automotive and commercial vehicle markets. Developed proprietary galvanized pad technology that addresses corrosion-related brake noise and premature failure in cold-weather and salt-belt markets—a growing differentiation point as OEMs focus on brake NVH.

Otto Zimmermann—Germany—Brake discs, drums, and pads exported to over 60 countries. A German rotor manufacturer with strong presence in the European aftermarket, recognized for their SPORT.Z line of drilled and slotted performance rotors.

Pacer Dynamics—Canada—Premium brake rotors with Geomet coating for passenger and commercial vehicles. A Canadian rotor specialist focused on corrosion-resistant brake discs, with Geomet-coated products gaining traction as OEMs and aftermarket buyers prioritize appearance and longevity.

PFC Brakes—USA—High-performance brake components for motorsports, aftermarket, and heavy-duty segments. Supplies carbon-metallic and ceramic friction materials to professional racing teams across multiple series, with technology that transfers into demanding street and heavy-duty applications.

Phoenix Friction Products—USA—Clutches and brakes serving OE replacement, performance, and heavy-duty fleet markets. Combines automotive brake friction with industrial clutch and brake products, serving a niche where these technologies overlap in heavy equipment and specialty vehicles.

PowerStop—USA—Performance brake kits for the aftermarket. A rapidly growing brand offering upgraded brake packages—including drilled/slotted rotors with ceramic pads—at accessible price points, with particular strength in e-commerce and direct-to-consumer sales.

ProTec Friction Group—USA—Advanced friction materials for commercial and industrial applications. Focused on heavy-duty and industrial braking segments where standard automotive friction formulations are inadequate for the extreme loads and environmental conditions involved.

PureForge—USA—Atomic-Forged rotors targeting law enforcement and public safety fleets. A specialty manufacturer whose forged rotor technology is designed for the extreme duty cycles of police pursuit vehicles and emergency response fleets, where repeated high-speed braking pushes conventional cast rotors past their limits.

Rane Brake Lining—India—Part of the Rane Group, produces brake linings, disc pads, clutch facings, and railway brake blocks. One of India's longest-established friction material manufacturers, with significant domestic market share built over decades of supplying Indian OEMs and the domestic aftermarket.

Rassini—Mexico—The largest vertically integrated brake disc manufacturer in the Americas. Controls the full production chain from iron melting through finished machined rotors, serving both OE programs for North American vehicle assembly plants and the aftermarket.

Remmen Brakes—Canada—Developing the Nonlinear Electro-Mechanical Brake (NEMB) system—a fully digital brake-by-wire technology. An emerging technology company developing next-generation EMB architecture, representing the type of new entrant that could disrupt traditional brake system suppliers.

Sangsin Brake—South Korea—Korea's leading brake pad producer with operations in China, India, the U.S., Mexico, and Brazil. One of the most globally diversified Asian friction manufacturers, supplying both Korean OEMs and international aftermarket channels under multiple brand names including Hardron and Hi-Q.

Scan-Pac Manufacturing—USA—Industrial friction and phenolic specialist, one of the early pioneers in developing non-asbestos friction materials. Now serving industrial and specialty braking applications where custom friction solutions are required for equipment ranging from elevators to heavy machinery.

Sebro—Germany—Brake disc and drum manufacturer and OE supplier to Porsche. A premium German manufacturer whose Porsche OE relationship signals the quality level of their casting and machining operations, with aftermarket products serving the European luxury vehicle segment.

Setfren Otomotiv—Turkey—Caliper repair kits, air brake system components, and disc brake pads for commercial vehicles. A Turkish manufacturer and exporter with its own branded product line, serving the heavy-duty aftermarket across Europe, the Middle East, and North Africa.

SGL Carbon—Germany—Carbon-fiber-reinforced silicon carbide (C/SiC) brake discs for high-performance and premium vehicle applications. A leading supplier of carbon-ceramic brake rotor material, producing the composite preforms and finished discs used on supercars and performance EVs. SGL partners with brake system integrators including Brembo and has expanded capacity as carbon-ceramic adoption grows beyond the supercar segment.

Shandong Gold Phoenix—China—The largest brake friction products manufacturer in China by production volume. A massive-scale operation supplying both the Chinese domestic market and global aftermarket, including private-label production for several well-known international brake brands.

Shandong Toget Brake System—China—Pneumatic disc brakes and calipers for trucks, buses, and trailers. A Chinese commercial vehicle brake specialist positioned to benefit from the accelerating adoption of air disc brakes in the Chinese heavy-duty market.

Shandong Xinyi Group—China—Large-scale brake pad and disc manufacturer. One of China's major brake component producers serving the domestic aftermarket and export channels, with production volume that makes them competitive on cost for global private-label programs.

SHW AG—Germany—Brake discs for premium European and American automakers. A German OE specialist whose customers include several luxury and performance brands, with particular expertise in ventilated and composite rotor designs for high-performance applications.

Sphere Brakes—USA—Developing the first patented sphere brake system using hemispherical pads against a spherical braking surface. An early-stage technology company pursuing a fundamentally different brake geometry that represents a speculative but potentially novel approach to brake system design.

Super Circle—India—Brake pads, brake shoes, brake linings, backing plates, and shims. A vertically integrated Indian manufacturer producing both finished friction products and component parts—backing plates and shims—that are supplied to other brake pad manufacturers globally. Over 65% of production is exported to more than 60 countries.

Sundaram Brake Linings—India—Friction materials including brake linings, disc pads, and clutch facings for commercial vehicles, passenger cars, tractors, railways, and two-wheelers. Part of the broader TVS Group heritage, Sundaram Brake Linings was established in 1976 and is a pioneer in asbestos-free friction material manufacturing in India. The company exports to more than 60 countries and serves both the Indian OEM market and global aftermarket channels.

Tenneco/DRiV—USA—Friction materials marketed under the Wagner and Ferodo brands, among others. One of the largest aftermarket brake portfolios globally, with Wagner as the dominant professional-installer brand in North America and Ferodo as its European counterpart.

TERREPOWER (formerly BBB Industries)—USA—The largest independent remanufacturer of brake calipers in North America, processing millions of cores annually. Positions remanufacturing as both a cost-effective and environmentally responsible alternative to new calipers, with a growing emphasis on sustainability as a business differentiator.

TMD Friction—Germany—Friction materials marketed under the Textar, Mintex, and Don brands. One of the world's largest independent friction material suppliers, with Textar as the leading OE brake pad brand in Europe and Mintex serving the European aftermarket. Acquired by Munich-based industrial group AEQUITA from Nisshinbo Holdings in late 2023.

TOMEX Brakes—Poland—Brake pads, brake shoes, sensors, and repair kits for passenger cars, light commercial vehicles, trucks, and buses, marketed under the TOMEX and BREXEL brands. One of Poland's most established brake manufacturers with over 40 years of production history, producing approximately 10 million brake pads annually.

TPB Auto—Vietnam—Brake drums for commercial vehicles. One of the emerging Southeast Asian foundry operations entering the brake component market, leveraging Vietnam's growing manufacturing base and cost competitiveness for export to aftermarket distributors.

Tribo—Ukraine—Friction materials manufacturer, the largest producer in Eastern Europe. Despite the challenges of operating during the ongoing conflict, Tribo has maintained production and continues to supply customers across Europe and the CIS region.

TSE Brakes—USA—Spring brake chambers and slack adjusters, part of the Marmon Group (Berkshire Hathaway). A specialist in commercial vehicle actuation components with deep expertise in spring brake chamber technology that is fundamental to heavy-duty fail-safe braking.

Walther Engineering (WEMC)—USA—Lightweight hubs, brake rotors, drums, and calipers for medium and heavy-duty trucks.

A commercial vehicle brake component manufacturer focused on weight reduction—a critical concern for fleets seeking to maximize payload within regulatory weight limits.

Wanxiang Qianchao—China—Brake systems, chassis components, and drivetrain products for both OE and aftermarket channels. One of China's largest and most diversified automotive component manufacturers, Wanxiang has grown from a domestic parts producer into a globally competitive supplier with OE relationships that extend beyond the Chinese market.

Webb Wheel Products—USA—Brake drums, rotors, hubs, and air disc brake calipers for Class 6–8 trucks. A long-established North American commercial vehicle brake component manufacturer with a particularly strong position in brake drums and wheel-end products for the heavy-duty segment.

Wilwood Engineering—USA—Performance brake kits for racing and high-performance street applications. A California-based manufacturer with deep motorsports roots, supplying complete brake systems to professional racing teams while offering a comprehensive catalog of performance upgrades for street vehicles.

Winhere Auto-Part Manufacturing—China—Brake rotors, drums, and pads. A large Chinese manufacturer with both OE and aftermarket business, including supply relationships with several North American aftermarket distributors and Tier 1 suppliers.

Yumak Otomotiv—Turkey—Air brake system products and pneumatic brake repair kits for heavy commercial vehicles. Founded in 1978 in Konya, Yumak is Turkey's leading manufacturer in the heavy commercial vehicle air brake segment, with a catalog of more than 10,000 precision spare parts and exports to over 100 countries across six continents.

ZF Group—Germany—Integrated chassis systems including brakes, steering, suspension, and driveline components. Dramatically expanded its braking portfolio through the 2020 acquisition of WABCO, the commercial vehicle braking specialist. Markets aftermarket brake products under the TRW brand.

Zhejiang Asia-Pacific Mechanical & Electronic (APG)—China—Complete brake system manufacturer producing disc and drum brake assemblies, vacuum boosters, ABS, ESC, calipers, and master cylinders. Founded in 1979 and publicly listed on the Shenzhen Stock Exchange, APG was the first Chinese company to independently develop and produce automotive hydraulic ABS.

HARDWARE & COMPONENT SPECIALISTS

Better Brake Parts—USA—Brake hardware kits for disc and drum applications. The market leader in brake hardware and installation kits in North America, providing the clips, springs, pins, and hardware that are essential to proper brake installation but often overlooked in the broader brake supply chain.

Cangzhou Xinlida Hardware Manufacturing Co. Ltd. (Xinlida)—China—Automotive brake components including guide pins, hardware kits, and related caliper hardware. A major Chinese manufacturer of the small but critical hardware components that brake pad and caliper manufacturers source for their assemblies.

Frenkit—Spain—Brake caliper repair kits, hardware, and components for the aftermarket and caliper rebuilders. A European specialist whose repair kits—seals, boots, pins, and hardware—support the caliper remanufacturing industry and independent repair shops across Europe.

Meneta—Denmark—Noise-damping brake shims, fine blanked backing plates, composite coil materials, and brake accessories. A Scandinavian manufacturer with specialized expertise in constrained-layer damping technology, supplying NVH solutions to brake pad manufacturers and OE Tier 1 suppliers worldwide.

NUADI Group—Spain—Brake pad backing plates, shims, and electronic wear sensors. A European component specialist supplying the precision-stamped backing plates and NVH-critical shim components that brake pad manufacturers integrate into their finished products.

NUCAP Brake Components—Canada—Brake system components and developer of the NRS galvanized brake pad technology. Their patented mechanical pad retention system and galvanized coating technology address the corrosion and delamination issues that affect conventional brake pads in harsh environments.

Preferred Automotive Components (PAC)—USA—Brake shims, wear sensors, and caliper hardware. A domestic supplier of the NVH solutions and electronic wear indicators that brake pad manufacturers and aftermarket brands incorporate into their finished products.

Trelleborg Sealing Solutions—Sweden—A world leader in noise-damping brake shims and insulators. A premium NVH component supplier whose constrained-layer damping shims are specified by major OE brake system suppliers globally, representing some of the most advanced noise mitigation technology in the industry.

Wolverine Advanced Materials—USA—Brake NVH solutions including the QUIETPATH brand of noise-damping shims and insulators. A domestic supplier competing in the brake noise mitigation space with proprietary constrained-layer damping materials for both OE and aftermarket applications.

BRAKE EMISSIONS & CAPTURE SYSTEMS

Mann+Hummel—Germany—Brake particulate capture systems. Best known as a global leader in filtration technology for engines and cabin air, Mann+Hummel has extended its expertise to brake dust capture as Euro 7 emissions regulations drive demand for at-source particulate mitigation.

Tallano Technologie—France—At-source brake particulate capture systems. A pioneer in brake dust capture technology, Tallano's TAMIC system uses a vacuum integrated into the brake caliper to capture particulates at the moment of friction, before they enter the air. Positioned for growth as Euro 7 regulations drive OEM adoption of brake emissions mitigation.

FOUNDRIES

Asama Coldwater Manufacturing (ACM)—USA—Gray iron brake rotors and drums. A Michigan-based foundry operation serving North American OE and aftermarket customers, originally established as a joint venture with Japanese partners bringing precision casting expertise to domestic production.

Draxton—Mexico—Iron and aluminum castings for brake systems, powertrain, and chassis applications. A leading North American brake casting supplier with vertically integrated operations, serving major OE programs and aftermarket customers from facilities in Mexico and Europe.

Fritz Winter—Germany—Cast iron brake discs and drums. One of Europe's major brake casting foundries, serving premium European automakers with gray iron rotors that meet the demanding metallurgical and dimensional specifications of German OEMs.

Kiriu Corporation—Japan—Brake discs and drums. A major Japanese brake foundry aligned with domestic OEMs through keiretsu-style relationships, with production facilities in Japan, China, and India serving the global operations of Japanese automakers.

MAT Foundry Group—UK—Cast iron brake discs, drums, caliper housings, and backing plates. A vertically integrated foundry group with operations across Europe, producing a wide range of brake castings for both OE and aftermarket customers from its multiple manufacturing sites.

Waupaca Foundry—USA—Gray iron and ductile iron castings including brake rotors, drums, calipers, and hubs. The largest iron foundry in North America, whose scale and metallurgical expertise make them a primary casting source for many of the continent's leading brake component manufacturers.

TESTING EQUIPMENT & SERVICES

Applus+IDIADA—Spain—Brake dynamometer testing, NVH analysis, brake durability testing, brake emissions measurement, and full vehicle brake development and engineering services. A global automotive engineering and testing company headquartered near Barcelona with over 30 years of experience and more than 3,300 engineers across 25 countries, whose dedicated braking systems department offers end-to-end capabilities from design and simulation through dynamometer and proving ground validation, including proprietary testing tools and a new GTR24-compliant brake particle emissions facility supporting Euro 7 compliance.

Bruker Nano Surfaces & Metrology—USA—Tribometers, mechanical testers, and surface metrology instruments for brake friction testing and particle emissions studies. Provides the precision measurement and tribological testing equipment that friction material developers use to characterize pad and rotor interactions at the material science level.

Greening, Inc.—USA—Brake dynamometers, friction testing equipment (including Chase machines), and ISO 17025-certified brake testing laboratory services. One of the industry's established providers of both testing equipment and contract testing services, with their Chase friction tester widely used for quality control screening of friction materials.

Horiba—Japan—Brake dynamometers, NVH testing systems, brake dust and particulate emissions measurement equipment, and comprehensive automotive test systems. A global leader in automotive testing instrumentation whose brake-specific equipment is found in OE and Tier 1 test laboratories worldwide.

Link Engineering—USA—World's largest brake testing laboratories, brake dynamometers, friction material testing (Chase, Krauss), NVH and emissions testing equipment, and vehicle testing services. The industry's most widely recognized name in brake testing, providing both the equipment that other companies use in their own labs and contract testing services for companies that outsource validation.

TecSA—Italy—Brake dynamometers for automotive and racing applications, along with brake testing laboratory services. An Italian testing specialist whose equipment and services serve both the European OE market and the motorsports community, with particular expertise in high-performance brake evaluation.

APPENDIX C: KEY INDUSTRY RESOURCES

STANDARDS ORGANIZATIONS

ISO (International Organization for Standardization)—Develops international standards including those for brake testing and performance.

NHTSA (National Highway Traffic Safety Administration)—U.S. agency that sets FMVSS safety standards including brake requirements. www.nhtsa.gov

The Insurance Institute for Highway Safety (IIHS)—Dedicated to reducing deaths, injuries and property damage from motor vehicle crashes through research and evaluation and through education of consumers, policymakers and safety professionals.

PUBLICATIONS

The BRAKE Report—Industry news, interviews, and analysis focused exclusively on the brake sector. www.thebrakereport.com

Brake Industry Compensation & Talent Report (Hagman Search)—Annual report with brake-specific salary ranges by function and experience level, talent scarcity ratings, and market analysis. Free download at www.hagmansearch.com.

SAE International Publications—Technical papers from the SAE Brake Colloquium and MOBILUS digital library provide peer-reviewed research on brake engineering topics. www.sae.org

Automotive News—Broad automotive industry coverage including supplier business news, OEM strategy, and regulatory developments that affect the brake sector. www.autonews.com

NAVIGATING THE CONFERENCE CIRCUIT

The brake industry's professional ecosystem is smaller and more relationship-driven than you might expect. The same faces appear at the same events year after year, and the informal conversations in hallways and at dinners often matter more than the formal presentations. Knowing which events to attend—and how to get value from them—is worth understanding.

Where to Invest Your Time

Not all events serve the same purpose. Here's how the major ones break down:

SAE Brake Colloquium & Exhibition is the premier technical conference for brake engineers. Organized by SAE International (founded in 1905 as the Society of Automotive Engineers, headquartered near Pittsburgh), it's held annually in the fall and draws 600–800 attendees and features peer-reviewed technical papers covering the latest research in friction materials, NVH, simulation, electronic braking, and testing methods. If you're in engineering or R&D, this is the one event you should prioritize. The technical content is genuinely substantive—papers presented here often represent years of research. The exhibition floor is smaller than trade shows but focused exclusively on brake-relevant technology and testing equipment. For career networking, it's where you'll meet the engineers you'll collaborate with (or compete against) throughout your career.

EuroBrake serves a similar function for the European market. Organized by FISITA (the international federation of automotive engineering societies, founded in Paris in 1948) since 2012, it draws over 1,000 delegates from 40+ countries and tends to emphasize topics where European regulations are driving innovation—brake particle emissions, sustainability, and advanced materials. If your company has European customers or operations, attending EuroBrake at least every other year keeps you current on regulatory developments that will eventually affect global markets. The technical program is strong, and the international mix of attendees provides perspective you won't get at U.S.-centric events.

Brake Weekend, hosted jointly by the Friction Materials Standards Institute (FMSI) and the Brake Manufacturers Council (BMC), has a different character. It's more commercially oriented, bringing together friction material manufacturers, aftermarket distributors, and data companies. The networking is excellent for aftermarket professionals, and the FMSI sessions on part-numbering standards and data management are practically useful for anyone in the aftermarket supply chain. It's also more intimate than the larger conferences, which makes it easier to build relationships.

AsiaBrake covers the rapidly growing Asian market, with particular focus on developments in India, China, and Southeast Asia. If your company is expanding in the region or competing against Asian suppliers, this event provides valuable market intelligence and networking opportunities that are hard to replicate from a distance.

AAPEX (Las Vegas, November) is the largest aftermarket trade show in North America, co-owned by the Auto Care Association and MEMA and held at The Venetian Expo. It's not brake-specific, but the brake category is well-represented among exhibitors. The value here is commercial—meeting aftermarket buyers, launching new product lines, understanding distribution trends. For engineers, it's less useful; for sales and marketing professionals, it's essential. It runs concurrently with the SEMA Show at the Las Vegas Convention Center—the world's largest automotive specialty products trade event, showcasing performance parts, custom accessories, and high-performance brake systems.

Automechanika (Frankfurt, Shanghai, Dubai, and other cities) is the international equivalent of AAPEX. The Frankfurt edition in particular draws global aftermarket players. If your company exports or competes internationally, Automechanika provides market exposure that AAPEX can't match.

HDAW (Heavy Duty Aftermarket Week) is the must-attend event for anyone in commercial vehicle brake aftermarket. Held each January at the Gaylord Texan Resort in Grapevine, Texas, it brings together heavy-duty distributors, manufacturers, and service providers. The format combines a traditional trade show with structured one-on-one meetings between manufacturers and distributors—an efficient way to conduct a year's worth of relationship-building in a few days.

PRI Show (Indiana Convention Center, Indianapolis, December) has served the motorsports community since 1988. If your company supplies racing applications or uses motorsports as a brand-building channel, PRI is where the racing industry congregates. It's also valuable for engineers interested in high-performance applications where braking technology is pushed to its limits.

Getting Value Beyond the Badge

Attending conferences is expensive—registration, travel, hotel, and the opportunity cost of time away from work. Here's how experienced industry professionals extract maximum value:

Prepare before you go. Review the technical program and flag sessions that are directly relevant to your current projects or interests. Identify specific people you want to meet—check the exhibitor list, speaker roster, and attendee list if available. Having a short list of objectives keeps your time focused.

Prioritize hallway conversations. The formal presentations are valuable, but the real intelligence often flows in the spaces between sessions. Coffee breaks, evening receptions, and dinners are where people speak more candidly about market trends, competitive dynamics, and career opportunities. Don't retreat to your hotel room between sessions.

Work the exhibition floor strategically. Walk the entire floor once to orient yourself, then return to booths where you have specific questions or relationship objectives. Booth staff at industry conferences tend to be technical people, not just sales representatives—they can have substantive conversations if you ask good questions.

Follow up within a week. Business cards collected at conferences have a short half-life. Send a brief email or LinkedIn message within a few days referencing your conversation. The relationships that last are the ones that continue beyond the event.

Consider presenting. Submitting a technical paper to SAE Brake Colloquium or EuroBrake, or participating in a panel discussion, elevates your professional visibility significantly. In an industry this small, being known as a subject matter expert in your area creates career opportunities that no amount of passive networking can match.

Industry Associations Worth Joining

Professional association membership provides access to technical resources, networking events, and standards development processes. The most relevant for brake industry professionals:

SAE International is the most broadly useful. Membership provides access to the MOBILUS digital library (an extensive archive of technical papers including decades of Brake Colloquium proceedings), discounts on conference registration, and involvement in standards committees where the technical rules of the industry are written. If you're an engineer, SAE membership pays for itself through access to technical literature alone.

FISITA membership connects you to the international engineering community and provides access to EuroBrake proceedings and other technical resources. It's particularly valuable if your work has a European dimension.

The Brake Manufacturers Council (BMC), representing the brake industry since 1973 and now an operating council within MEMA's Aftermarket group, serves companies rather than individuals. But if your employer is a member, take advantage of the networking and market intelligence it provides. BMC's focus on aftermarket brake business makes it especially relevant for commercial professionals.

FMSI (Friction Materials Standards Institute) membership is valuable for anyone involved in aftermarket part numbering, cataloging, or data management. Founded in 1948 by North American aftermarket friction manufacturers, the organization's D-Number system is the common language of the aftermarket, and understanding how it works—and participating in its development—is practically useful.

MEMA (Motor & Equipment Manufacturers Association) provides advocacy, benchmarking data, and networking at the broader automotive supplier level. Its value is more strategic than tactical—useful for executives and business development professionals who need to understand industry-wide trends.

A Note on Professional Development

The brake industry doesn't have a formal certification program the way some fields do, but there are meaningful ways to signal and develop expertise. Presenting at conferences, publishing technical papers, participating in SAE standards committees, and contributing to industry working groups (like those developing Euro 7 test methods or updated FMVSS requirements) all build professional credibility in ways that internal company work alone cannot. The industry is small enough that your reputation precedes you—invest in building one that opens doors.

APPENDIX D: CAREER TOOLS

The following tools distill the career guidance from Part V into actionable frameworks. Use them as starting points and adapt to your specific situation.

COMPANY EVALUATION SCORECARD

Rate each factor 1–5. A score below 3 in any category warrants serious consideration before accepting an offer. A total score below 18 suggests proceeding with caution.

Financial Health (____/5): Revenue trend, profitability, debt levels, customer concentration, recent investments. Check public filings, news coverage, and industry reputation. Ask yourself: Are they growing or just surviving?

Customer Relationships (____/5): Key OE and aftermarket partnerships, recent program wins or losses, customer diversification. Who are their top three customers? What percentage of revenue comes from each?

Technology Position (____/5): R&D investment, patent portfolio, product roadmap alignment with industry trends (EV, EMB, Euro 7). Are they leading, following, or scrambling to catch up?

Manufacturing & Quality (____/5): Plant conditions, equipment age, quality certifications, customer scorecard performance. What are their PPM rates and delivery metrics? Any recent escapes or recalls?

Culture & Leadership (____/5): Turnover rates, employee reviews, leadership tenure and stability, promotion-from-within patterns. Talk to current and former employees. Trust your interview impressions.

Career Trajectory (____/5): Role clarity, defined growth paths, training investment, succession planning. Will this position prepare you for your next career move?

Total Score: ____/30

INTERVIEW PREPARATION CHECKLIST

Complete this checklist before every interview. Thorough preparation is what separates good candidates from great ones.

Research (Complete First)

- ☐ Review company website, recent press releases, and LinkedIn page
- ☐ Identify key products, major customers, and recent program wins or losses
- ☐ Assess financial health through public filings, news coverage, and industry sources
- ☐ Research interviewers' backgrounds via LinkedIn, publications, and patents
- ☐ Understand recent industry trends specifically affecting this company

Your Stories (Prepare 5–7 Examples)

- ☐ A problem you diagnosed and solved, demonstrating technical depth
- ☐ A process you improved or cost you reduced, with quantified impact
- ☐ A difficult situation with a customer or colleague you navigated successfully
- ☐ A failure you experienced and what you learned from it

- ☐ A cross-functional project you led or contributed to significantly
- ☐ Your specific reasons for interest in this company and role

Role-Specific Preparation

Engineering: Prepare an NVH case study, DFMEA example, testing methodology explanation, and material selection rationale.

Quality: Prepare an escape recovery story, audit experience summary, supplier development example, and PPM improvement case.

Sales: Prepare stories about deals won and lost, a customer relationship recovery, and territory or account growth examples.

Operations: Prepare efficiency improvements with metrics, team development examples, and continuous improvement initiatives.

Questions to Ask Them

- ☐ What does success look like in this role at six months and twelve months?
- ☐ What is the biggest challenge facing this team right now?
- ☐ How does this role interact with other departments?
- ☐ What happened to the previous person in this position?
- ☐ What is your timeline for making a hiring decision?

CANDIDATE EVALUATION MATRIX

For Hiring Managers. Score each candidate 1–5 on these dimensions using behavioral evidence rather than general impressions. Compare all candidates against the same criteria for consistency.

Technical Competence (Table Stakes)

Domain Knowledge (____/5): Does the candidate understand brake systems, materials, testing, or their functional area at the depth required for this role?

Problem-Solving Approach (____/5): Does the candidate describe how they diagnosed problems, not just that they solved them? Look for methodology.

Technical Communication (____/5): Can the candidate explain complex concepts clearly? Do they adjust their communication to match their audience?

Differentiators (What Separates Good from Great)

Customer Orientation (____/5): Does the candidate discuss customer impact? Do they understand the customer's customer and the broader value chain?

Ownership Mentality (____/5): Does the candidate describe problems they personally owned, or problems they merely observed? Listen for 'I' on individual contributions versus vague 'we' statements.

Learning Agility (____/5): How has the candidate handled unfamiliar situations? Do they actively seek feedback? Can they demonstrate growth over their career?

Cultural Fit (____/5): Will this candidate thrive in your environment? Do their values align with your organization? How do they speak about previous employers?

Red Flags

- ✖ Blames others for failures while taking personal credit for team successes
- ✖ Cannot provide specific examples; speaks only in generalities
- ✖ Disparages previous employers (they will eventually do the same to you)
- ✖ Asks no questions about the role, team, company, or challenges
- ✖ Oversells credentials; specific claims do not match résumé details
- ✖ Demonstrates poor listening; answers questions that were not asked

Final Assessment
Total Score: ____/35

Recommendation:
☐ Strong Hire ☐ Hire ☐ Needs Discussion ☐ No Hire

Key Strengths:

Key Concerns:

Follow-up Questions for Next Round:

APPENDIX E: TECHNICAL DEEP DIVES

The following sections provide detailed technical content referenced throughout the main chapters. They're organized by source chapter so you can easily find the deep dive that corresponds to what you're reading. This material is designed for engineers, technical professionals, and anyone who wants the full picture behind the practical summaries in the main text.

FROM CHAPTER 2: HOW BRAKES ACTUALLY WORK

Rotor Metallurgy

Most brake rotors are gray cast iron—not because it's the best possible material, but because it offers an exceptional balance of properties at acceptable cost.

Gray iron gets its name from the gray appearance of its fracture surface, caused by graphite flakes distributed throughout the iron matrix. These graphite flakes are actually crucial to brake performance: they provide lubrication at the pad-rotor interface, help conduct heat, and give the iron good damping characteristics that reduce noise.

The specific composition matters enormously. Carbon content typically runs 3.4–3.8%—too little carbon and the iron becomes too hard and prone to cracking; too much and it becomes weak. Silicon content (usually 1.8–2.4%) promotes graphite formation. Small amounts of manganese, copper, and other elements fine-tune properties.

"High-carbon" rotors have become a marketing term in the aftermarket, but the designation can be misleading. What matters isn't just carbon content but the size, shape, and distribution of graphite flakes—characteristics that depend on the entire casting and cooling process, not just chemistry.

Phenolic vs. Steel Pistons

The piston that pushes the pad might seem like a minor component, but material selection matters:

Steel pistons are strong, stiff, and durable. They conduct heat well—which can be either advantage or disadvantage. In a well-designed system, a steel piston helps dissipate heat. In a marginal system, that thermal conductivity can transfer too much heat to the brake fluid, risking boil.

Phenolic (plastic composite) pistons are common in modern calipers. They don't conduct heat as readily as steel, helping insulate the brake fluid from rotor heat. They're also lighter. The material is more than adequate for the compressive loads involved—pistons primarily push, they don't need to resist bending or twisting forces.

Brake Fluid Specifications

Brake fluid is engineered to remain stable across an enormous temperature range while transmitting force without compression. The DOT (Department of Transportation) ratings specify minimum performance:

DOT 3 is the baseline glycol-ether fluid with a minimum dry boiling point of 401°F and wet (moisture-contaminated) boiling point of 284°F. It's adequate for normal passenger car use.

DOT 4 raises the minimums to 446°F dry and 311°F wet, providing better margin for performance driving or heavy-duty use.

DOT 5.1 (not to be confused with DOT 5) further increases boiling points to 500°F dry and 356°F wet while remaining glycol-based and compatible with DOT 3 and 4.

DOT 5 is silicone-based rather than glycol-based. It doesn't absorb water, which sounds like an advantage—but moisture that enters the system pools rather than dispersing, potentially causing localized corrosion and ice formation. DOT 5 is also compressible under certain conditions and is not compatible with ABS systems. Its primary use is in vehicles that sit for extended periods (military, collector cars) where moisture absorption would be problematic.

Why Boiling Points Matter

The glycol-based fluids (DOT 3, 4, 5.1) are hygroscopic—they absorb moisture from the atmosphere. Water contamination is inevitable over time as moisture migrates through hoses and seals.

This matters because water dramatically lowers boiling point. A system filled with fresh DOT 4 fluid might boast a 446°F boiling point—but after two years of moisture absorption, that same fluid might boil at 311°F or lower.

When brake fluid boils, it creates vapor bubbles that compress under pressure. Instead of transmitting force to the calipers, the pedal simply compresses gas. The result is a spongy pedal that sinks toward the floor without producing adequate braking—a terrifying experience that has caused countless accidents.

This is why brake fluid requires periodic replacement regardless of apparent condition, and why track drivers often flush systems immediately before events.

Brake Torque Calculation

The torque that slows your wheel equals: Clamping Force × Coefficient of Friction × Effective Radius

Clamping force comes from hydraulic pressure acting on piston area. More pressure or larger pistons mean more force.

Coefficient of friction depends on the pad material interacting with the rotor surface. It varies with temperature, pressure, and speed—which is why brake performance changes under different conditions.

Effective radius is the distance from the wheel center to where braking force acts—essentially, the middle of the pad contact area. Larger rotors have larger effective radii, increasing torque for the same clamping force. This is why "big brake kits" improve braking: not because they generate more heat (though they can dissipate more), but because larger rotors multiply the effectiveness of available clamping force.

FROM CHAPTER 8: THE MATERIAL WARS

Binders: The Foundation

Phenolic resin is the most common binder in brake friction materials. This thermosetting polymer forms the matrix that holds all other ingredients together.

During pad manufacturing, uncured phenolic resin is mixed with other ingredients, pressed into shape, and then heat-cured. The curing process cross-links polymer chains, creating a rigid structure that won't melt under braking heat (though it will eventually degrade at extreme temperatures).

Binder content must be carefully balanced. Too little binder and the pad lacks strength—it may crack under stress or wear too quickly. Too much binder and friction properties suffer as the pad becomes too "slippery."

Some formulations use modified phenolic resins—incorporating rubber or other modifiers—to improve specific properties like noise damping or cold performance.

Reinforcing Fibers: The Skeleton

Fibers provide structural reinforcement, like rebar in concrete. Different fiber types contribute different properties:

Steel fibers offer strength, heat conductivity, and consistent friction. They're a staple of semi-metallic formulations. The downside: they can be aggressive on rotors and contribute to noise.

Aramid fibers (synthetic fibers like Kevlar) provide excellent strength with lower weight and reduced rotor wear. They're common in premium NAO formulations but add cost.

Ceramic fibers (various compositions) handle heat well and provide stable friction. They're key ingredients in ceramic pad formulations.

Mineral fibers (like processed mineral fiber) offer heat resistance and reinforcement at lower cost than exotic alternatives. Environmental concerns have reduced some mineral fiber use.

Glass fibers add strength and help control friction. They're sometimes used in combination with other fiber types.

Copper fibers historically provided excellent heat transfer and friction stability—but environmental regulations have largely eliminated copper from new formulations, as discussed elsewhere.

The fiber mix significantly affects pad character. A high-steel formulation feels different from high-aramid—they have different friction curves, wear patterns, and noise tendencies.

Friction Modifiers: Tuning the Grip

Friction modifiers adjust the coefficient of friction—how much grip the pad provides at different temperatures and pressures.

Graphite is ubiquitous in friction formulations. This form of carbon provides lubrication (reducing noise and stick-slip), improves heat dissipation, and helps stabilize friction at high temperatures. Graphite content can range from a few percent to over 10%, depending on the formulation.

Metal sulfides (antimony trisulfide, zinc sulfide, tin sulfide) are common friction modifiers that provide stable friction characteristics, especially at elevated temperatures. Antimony compounds have faced environmental scrutiny; alternatives are increasingly used.

Molybdenum disulfide acts as a high-temperature lubricant, reducing friction at extreme temperatures where other modifiers degrade.

Cashew dust (yes, from cashew nut shells) is a traditional friction modifier providing consistent friction and some binding properties. It's been used for decades and remains common.

Rubber particles improve flexibility and damping, helping control noise. They contribute to cold-weather grip but degrade at high temperatures.

Abrasives: Controlled Aggression

Abrasives might seem counterintuitive in a friction material—why add something that causes wear? But controlled abrasiveness serves important functions:

Aluminum oxide and silicon carbide are hard particles that abrade the rotor surface, maintaining a clean interface for consistent friction. Without some abrasive action, rotors would glaze over, reducing friction.

These particles also directly contribute to friction. The scratching action of abrasive particles against the rotor generates stopping force.

But abrasive content must be carefully controlled. Too much abrasive and rotor wear becomes excessive. Different abrasive types have different effects on both friction and wear.

Fillers: Bulk and Balance

Fillers add volume and adjust physical properties without dramatically changing friction behavior:

Barium sulfate (barytes) is a common filler that adds density and thermal mass without negative effects. It's relatively inert and inexpensive.

Calcium carbonate (limestone) serves similar purposes and helps moderate friction at high temperatures.

Mica contributes to heat resistance and helps control pad expansion under heat.

These fillers might seem like cheap bulk material—and cost reduction is one benefit—but they also allow formulators to fine-tune density, thermal properties, and manufacturing behavior.

Mode Coupling: The Technical Explanation

Engineers describe brake squeal as a mode coupling instability. The brake system has multiple vibration modes—ways the components can oscillate. Under certain conditions, two modes can couple together, exchanging energy in a way that reinforces vibration rather than damping it.

Imagine the rotor vibrating in a "waving" mode while the pads vibrate in a "rocking" mode. If these modes have similar frequencies and the right phase relationship, energy from braking can continuously feed into the vibration, creating sustained squeal.

This coupling depends on many factors: friction coefficient, contact pressure, temperature, component geometry, and material properties.

Small changes in any of these can push the system from quiet operation into squeal—or vice versa.

Iron Metallurgy Details

The gray iron used for brake rotors is specifically formulated for the application. Carbon content (typically 3.4–3.8%) determines graphite formation. Silicon (1.8–2.4%) promotes graphite flake formation and adds some strength. Manganese, phosphorus, sulfur, and other trace elements fine-tune properties.

The graphite flake structure is critical. During solidification, carbon precipitates as graphite flakes distributed throughout the iron matrix. These flakes provide:

Lubrication at the friction interface, reducing stick-slip and noise.

Thermal conductivity, helping distribute heat throughout the rotor.

Damping characteristics, absorbing vibration that could cause noise.

Machinability, making the rotors easier to manufacture.

Different graphite morphologies (flake size, distribution, orientation) significantly affect rotor behavior. Controlling these characteristics requires precise control of chemistry and cooling rates during casting.

ACKNOWLEDGMENTS

First and above all, I thank God. Every good gift comes from above, and He's been generous with me. His grace and guidance are behind this book, the businesses, and the relationships that made it all happen. I'm thankful for all of it.

To my wife, Anne—thank you for putting up with me through this process and encouraging me every step of the way. To my sons, Nolan and Bennett—you two are always in my corner, and that means more than you know. I'm a lucky man.

A special thank you to Greg Vyletel, whose technical review made this a better book. Greg brought decades of brake engineering knowledge to many chapters—catching errors, sharpening the details, and adding insights that only someone who has lived this work could provide.

My sincere thanks to Roy Link for contributing the foreword to this book. Roy's willingness to lend his name and perspective to this project means a great deal, and his foreword is the perfect way to open the book.

A book like this doesn't come from one person's knowledge. It's the result of hundreds of conversations over more than a decade—with engineers, executives, and professionals across the brake industry who were generous enough to share what they know. Through interviews for *The BRAKE Report*, recruiting conversations, factory tours, and conferences, you taught me how this industry really works. I've done my best to honor that trust here.

To the hiring managers who have partnered with Hagman Search—your candid explanations of what you look for directly shaped the career chapters. To the candidates I've worked with over the years—your questions and experiences taught me what people actually need to know about building a career in this industry. A lot of this book started as answers to questions you asked me.

To the readers of *The BRAKE Report*—your engagement over the years confirmed there was an appetite for something like this. And to the industry associations and conference organizers who have welcomed me into their events—those gatherings are where many of these relationships started.

Any errors in this book are mine alone. The brake industry is complex and constantly evolving, and no single person can know all of it. I encourage readers to verify critical details through additional sources.

To everyone entering or already working in this industry—I hope this book helps. It's an industry that rewards deep expertise and long-term commitment, and I wish you success.

INDEX

U–V

W–Z